D0048619

THIS
ABUNDANT
LAND

Edited by John A. Prestbo

DOW JONES BOOKS
PRINCETON, N.J.

Introduction

People are rediscovering U.S. agriculture as a major factor in the economic, social and political well-being of not only America, but of the world as well.

For much of this century people have been preoccupied with rapid industrialization and urbanization. Now, a whole generation has grown up during a period when there was always enough—more than enough—food to eat. It was easy to take agriculture for granted.

As recently as early 1972, the nagging "farm problem" was how to control the potent productive capacity of U.S. agriculture. The government paid farmers not to plant certain crops, but still surpluses piled up. Food prices were relatively reasonable, but taxpayers were burdened with billions of dollars in subsidy payments, which many farmers depended upon to stay in business.

All that changed quickly in the summer of 1972 when the size of the Soviet Union's massive purchases of U.S. grain became known. The U.S. was besieged with orders from Europe, Japan and other countries that were desperate for grain and other farm products because of crop failures around the world. Within a year, the U.S. practically ran out of soybeans, and the government curtailed exports temporarily. Partly because of price controls, other foodstuffs came into short supply, too, and many Americans for the first time in their lives encountered empty shelves in their supermarkets. Food prices rose dizzyingly.

Suddenly, the production of food and fibers hit home for millions of families in America and abroad who depend on U.S. agriculture—and who know almost nothing about it.

It is surprising to many people, for instance, that agriculture is still the number one industry in the U.S. By some estimates, one of every three jobs in the country is involved with or dependent on the business of putting food on our tables and providing natural fibers for our looms. Tremendous amounts of money are spent annually for new equipment, fertilizer, and other supplies. This has a ripple effect throughout the economy, stretching back to such basic industries as steel, rubber, petroleum and chemicals.

Surprising, too, are the many facets of agriculture. This book includes tales of grain farmers, cattle ranchers, sheepmen, and a variety of others, including those who grow everything organically in compost and those who are growing crops without any soil at all. You'll also read about the vast, diversified world of agribusiness—the network of corporations and cooperatives that supply farmers and ranchers with what they need, that buy their products and that, in a few cases, do some farming themselves.

Agriculture and politics have long been entwined. A section of this book, therefore, describes how farmers' political power has been eroded by urbanization, but nonetheless still remains potent. Another section elaborates on the increasingly frequent clashes between the agriculture industry and environmentalists and of efforts to clean up farm pollution. You will read, too, of researchers' continuing efforts to boost agricultural productivity—already by far the highest in the world and in history—and to develop new crops.

You will be introduced here to successful farmers and struggling farmers, the bosses and the bossed, those who are seeking to expand and those who are calling it quits. You will meet a young couple trying to become self-sufficient, pioneer style, on a plot in Maine, and a family that has farmed in New Hampshire for 11 generations.

All this makes interesting reading, to be sure, because the people here are involved in unusual ways of making a living and, in most cases, a distinctive way of life. The hopes, the risks, the successes, the setbacks—all are part of the quiet drama which goes on, mostly unnoticed, in America's farm communities.

But the stories also illuminate the economics and the sociology of agriculture in America today. They detail the impact of droughts, floods, and politics on supermarket prices. They discuss fertilizers and chemicals which increase productivity—but may be hazardous to your health. They report on research to improve crops and create new ones. (One innovator, for instance, has come up with a "burpless" cucumber. One company discovered that air conditioning improves a hog's sex life—and thus increases the production of pigs).

The material in this book was gathered and reported by staff writers of The Wall Street Journal which in recent years has made a sustained effort to spot and report on trends in agriculture. In their special way of reporting and writing, the staff and its editors have found imaginative ways of separating the chaff from the wheat—and making the results interesting as well as useful.

A personal note: Nearly all the people in agriculture share some common characteristics—notably, a deep and abiding love of the land and an amazing capacity for hard work.

My introduction to agriculture came from my late grandfather, John Schol, who ran a grain farm in eastern North Dakota. I visited him almost every summer when I was a little boy. I remember being awakened even before the gray dawn of early July by the sound of him starting his old Chevrolet (which he jokingly dubbed "the Cadillac") and chugging off to chores that often weren't done until dark. I remember sitting with him at his kitchen table as he listened intently to the radio for weather reports and, with a resigned shake of his head, to the latest crop prices. And I remember deciding that farming was not for me.

So it came as something less than a pleasant surprise when, as a young reporter in the Chicago bureau of The Wall Street Journal, I was assigned to the agricultural beat. I couldn't have been more mistaken. In the seven years I covered that beat, 1966-1973, I found agriculture to be one of the most exciting and truly interesting of man's many endeavors. And the men and women of agriculture are among the most refreshing people I've encountered.

I'll never forget going to Dodge City, Kansas, one April to report on a drought that was withering the winter-wheat crop. During the course of my stay I became acquainted with one farmer in particular, and he invited me to his home for "lunch" (which turned out to be their big meal of the day and which consisted of beef, ham, chicken, several kinds of vegetables and potatoes, and dessert, all of which had been prepared by his wife that morning). Afterwards, we chatted about a variety of things, including the economics of farming. To back up a point he was making, my new friend— whom I'd met only a day earlier and who knew me to be a nosy newspaper reporter—dug out his income-tax returns for the previous four years and went over them with me. I was astonished—and touched—by his trusting, yet self-assured openness.

Now that I no longer visit the fields, but instead sit with a pencil in my hand at a desk in Manhattan, I occasionally re-read the stories of farm people—an attempt, I think, to keep in touch with them. I hope that in these pages you will catch a memorable glimpse of the remarkable people whose labors have made U.S. agriculture the envy of the world. It is to these people who work this abundant land that this book is dedicated.

—JOHN A. PRESTBO
Editor

Contents

CONTENTS

THIS
ABUNDANT
LAND

Part One

THE PEOPLE WHO WORK THE LAND

The people who work the land are conservative, for the most part, and many of them have a strong sense of tradition. Yet they are hunch-playing, risk-taking gamblers by instinct and experience. Many of them are wary and skeptical of change forced upon them, but at the same time they are often more innovative and experimental than most business executives would ever dare to be. They sometimes seem reticent and reserved, especially on first meeting, but just as frequently their emotions are manifest and their plainspokenness is eloquent. Most of them love the land and their work very deeply, and that is as much a national asset as the fertile soil.

Up on the Farm

"The cab isn't air-conditioned, and it hasn't got stereo, but we're sure going to use it anyway," says Jerry King as he watches his new bright-orange tractor being delivered to his farm near the west-central Illinois town of Victoria.

The $19,000 tractor is the biggest and the most expensive that Mr. King has bought in 14 years of farming, but he thinks it's worth the price because it will plow six acres an hour, compared to his three-year-old tractor's four acres. Power and capacity are more important than comfort to Mr. King, a short, wiry man of 36 who perches on the edge of his chair as he speaks intently about farming.

"It's a whole new way of farming from when I started," he says. "The Main Street businessman has had to change with the times, and so does the family farmer. We've got to move ahead."

That means expansion, as it has for most farmers since World War II—but with a big difference. For many years, U.S. farmers had to struggle to expand their individual operations against a national policy geared to restraining production. Now, because of shrinking supplies and growing demand for food and fiber around the world, the farmers' natural inclination to grow is being spurred by higher prices for their products and a farm program that is bringing millions of retired acres back into production.

So prosperity hits the Farm Belt in a special way. Farmers spend some of their money for color TVs, winter vacations in Hawaii and new cars, but most of it is being plowed back into the farms themselves. Though inflation soaks up a good part of the reinvestment, a significant portion is directed toward increasing efficiency and production.

Mr. King exemplifies many of these trends under way in modern American agriculture. He raises and sells 4,000 hogs a year, and he feeds them mostly with corn that he grows on 600 of his 780 acres. He made an average profit of $10 a hog in 1973, compared with about $5 in 1972.

Some of this added profit went to an extensive remodeling of his home, but Mr. King spent most of it on his hogs. Besides buying a new tractor and other equipment, he hired a second full-time employe and is planning to build a $70,000 hog house, which would be his third. He figures the house, measuring 33 feet by 232 feet, will boost his output by 1,800 hogs annually without adding more labor and will return about three times the amount he'd earn by depositing his profits in the nearby First National Bank of Galva.

When Mr. King started farming in 1960 with the pigs from 15 sows, "we carried everything by hand"—from feed to manure. His big new hog houses, by contrast, are models of U.S. agricultural efficiency. The animals stay inside them from the time they're born until they are sent to market six months and 200 pounds later. They eat a corn-based feed that is scientifically formulated for nutrition and weight gain, and they drink water from automatic faucets in their pens. Manure is collected beneath the slotted floors of the pens and is flushed outside for use as fertilizer.

Mr. King is borrowing heavily to finance his expansion, which is another trait increasingly common among modern farmers. His debt load isn't excessive, as long as he steadily sells hogs to keep cash coming in. To this end, he staggers pigs' births so that he has some hogs to sell each week. But he concedes the risks are still big.

"I worry about becoming so dependent on our present high cash turnover that we can't cut back if times get hard," he says.

The credit standing of Mr. King and other big-borrowing farmers is dependent largely on their net worth, which is being bolstered considerably by fast-rising land values. Prices of farmland rose 13% in the year ended March 31, 1973, and they have increased by 15% or more since then, farm realtors estimate.

But rising land prices work against farmers such as Morris Greenley, who raises 1,400 acres of corn and soybeans on the gently rolling land south of Independence, Iowa. He rented 600 of these acres for about $50 each, but similar agreements up for renegotiation on nearby farms boosted the rates to $65 to $75 an acre in the summer of 1973.

"I'd like to work about 20% more ground," says Mr. Greenley, a graying, husky man of 39, as he taps out figures on an electronic calculator on the office desk in his four-year-old farm home. "I can handle that much more with equipment I've got, and it would spread out my fixed costs more." He makes a 10% return on his investment of about $174,000, plus a "decent" salary, but he figures that he needs to expand to offset the erosion of steadily climbing costs.

What he doesn't know is how soon he can find some land to rent that is close enough for him to farm efficiently. Most of the acreage he has picked up has come from farms whose owners quit during leaner years, either because of age or to work at higher-paying jobs in towns. But the farm boom is slowing and in some areas even reversing the migration from the land.

Lloyd Phelps Jr. of Cedar Rapids, Iowa, is one of those who has changed his mind about farming as a career. He grew up on a farm, but when he graduated from high school he took a job as a mechanic in FMC Corp.'s Link-Belt machinery plant in Cedar Rapids because he didn't think he could make the living he wanted as a farmer. Now he spends his spare time helping on his father's 300-acre farm and raising 200 acres of corn and soybeans for himself on rented land.

"In about five years, I'll probably be farming full time," he said one afternoon in September 1973 as he relaxed after the 6 a.m.-3:30 p.m. shift with a cup of coffee and a big piece of cake at his kitchen table. By then, he will have been at Link-Belt 15 years and his stake in its employe-investment plan will be fully vested; also, by then, he figures, "my father probably will have retired."

Higher corn and soybean prices were instrumental in his decision to return to farming, he says. With overtime, he makes about $12,000 from Link-Belt, and in 1973 he netted a similar amount from his farming.

"Farming has bought us a lot of nice things we couldn't have had otherwise," he says. "But it's pretty hectic. I'm working 65 hours a week at the plant. I get the field work done after supper and on weekends, and I split my vacation between the spring and fall busy periods." His wife, Sandra, and their four children say they don't mind his fast pace.

While the farm boom is slowing the decline in farms and farmers, it might be hastening the disappearance of small towns that once dotted the countryside. Earl Heady, an agricultural economist at Iowa State University in Ames, says that as family farms get bigger by substituting capital for labor the farmers tend to go to county seats or larger towns for the loans, equipment and supplies that they need in greater quantities. Without this economic base, the tiny farm towns quickly dry up.

That trend is already marked in the Deep South, says G. C. Cortright, a cotton grower in Rolling Fork, Miss. "Thirty or 40 years ago, when we had maybe a hundred families living on our plantation, there used to be a black family on every 20 acres and small towns five or seven miles apart," he recalls. "Now, we've got less than two dozen employes. The families are gone, and those little towns are dying."

Mr. Cortright is the third generation of his family to manage the plantation of more than 1,200 flat, sandy-loam acres of rich Mississippi Delta land. His predecessors oversaw black men and women doing the backbreaking work of hoeing the rows of young plants and later picking the white tufts and stuffing them into bags. By contrast, Mr. Cortright sprays chemicals to control weeds and bugs, and sends four red-and-white International Harvester cotton pickers rumbling across the fields. Each of these giant machines can pick 21,000 pounds of cotton a day, compared with 400 for a top-notch field hand.

"The machinery and the sophisticated use of chemicals are two big changes to come to cotton farming," he says. "Now the competition of soybeans for cotton ground seems to be another." Because of high prices, soybeans and wheat are encroaching on land where cotton once was king. "We can get a double crop of wheat and soybeans here each year instead of a single crop of cotton, so unless cotton prices stay high a lot of cotton farmers are going to switch over," Mr. Cortright says.

Spring floods prevented many farmers, including Mr. Cortright, from planting as much cotton as they had intended in 1973, so instead they put in soybeans, which can be seeded much later in the season. As a result, the 1973 cotton crop was smaller than expected, and cotton prices—which hit 99 cents a pound on the

New York futures market—were at their highest since the Civil War.

Some cotton farmers were flooded out entirely, though, while in other parts of the country growers of fruit and vegetables were hit by frost, hail and other calamities.

Dairymen, in particular, were hard hit. The price that Wisconsin dairy farmers get for their milk rose 31% from September 1972 to September 1973, but the cost of corn and soybeans that make up the bulk of their feeds has more than doubled in that period.

As a result, milk production dropped, many dairy cows sold for beef, and many farmers quit.

"The costs keep going up," says Wilfrid Turba, who milks a 40-cow herd on 230 hilly acres northeast of Fond du Lac, Wis. "A lot of us have had to use more home-grown feed for our cows. The cows don't produce as much milk, but it's the only way we can afford to feed them. When you compare the way grain prices have risen with the way milk prices haven't, you've got some idea why anyone with an inclination to get out does it."

Mr. Turba, a muscular, soft-spoken man, isn't so inclined, but he indicates that he is resigned to the possibility of sharply reduced income lasting for several more years. One of the few bright spots is that one of his nine children, Michael, is thinking of joining him on the farm, which Mr. Turba's great-great-grandfather first settled 125 years ago.

"When I started 20 years ago, it was common for fathers to sell their sons a farm and knock a couple thousand dollars off the price as a wedding present," he says. "Nobody has done that in a long time; land costs $500 an acre instead of $100. If Mike comes in, we'll expand the herd and he'll be a partner. It's about the only way he could afford to start."

—GENE MEYER

Making Ends Meet

Charlie Stephenson grew a bumper corn crop on his 420 rented acres in 1971. When fall came, it took him and his wife, Ruth, a full month of exhausting 16-hour days to harvest the crop.

Each day, Charlie and Ruth rose before dawn and headed for the fields, leaving their children, Bob and Sue, to get themselves off to school. Ruth, her hair protected from the dust by a bandana, steered the big green combine across the fields while Charlie dried the corn and stored it. There was so much of it that, even with a new bin, the Stephensons ran out of places to put it. It looked certain to be a good year.

Then Charlie tried to sell the corn—and learned that, financially, there was no way he could reap what he had sowed. Corn farmers all over the nation, it seemed, harvested a bumper crop that year. That depressed corn prices to about two-thirds of 1970 prices, and in the end Charlie Stephenson's profit came to about $6,000, or nearly $550 less than a smaller harvest had brought him the previous year.

"Sometimes I feel like I'm beating my head against a wall," said Charlie, a sturdy, ruddy-faced man who that year was 35 years old. "I work harder and make less. We'll just have to tighten our belts again."

Charlie's plight was shared by most of the nation's million farmers. Soaring costs of everything except the products they sell had squeezed farmers for years. But the 1971 pinch was particularly acute. "Farmers are the only ones in the country who get stuck for being too efficient," says Mardy Myers, a farm-income specialist with the U.S. Department of Agriculture.

Indeed, it appeared to more and more people that the small farmer like Charlie Stephenson was in danger of disappearing altogether from the American landscape. From California to New York, enormous corporate farms, consisting of thousands of acres of the best land and backed by enormous amounts of capital, have replaced farmers with a few hundred acres of land, no capital and plenty of debt. Three firms—Purex, United Brands and Bud Antle—were producing so much of the lettuce sold in America that their hold on the market sparked a rare agricultural antitrust investigation by the Federal Trade Commission. Earl Butz, President Nixon's Secretary of Agriculture, was almost rejected by the Senate after he testified in favor of corporate farming and revealed that he was a director of such firms as Ralston Purina and Stokely-Van Camp.

Ironically, the demise of the farmer that so dismays some in the government is in large part a result of government policy over the years. It was the biggest farms that received the biggest subsidies and the most government-supplied irrigation water. Likewise, federal tax laws favored the corporation and the investor who farms on the side in search of a tax write-off over the family man who farms full time in search of a living. These laws and farm programs have been modified since then, and in any case concern about the plight of farmers diminished during the farm boom of 1972-73. If hard times return, though, some experts think the family-farm struggles of the Stephensons in 1971 could again become typical.

Charlie and Ruth Stephenson found in late 1971 that they couldn't make ends meet with the money they made from farming. Ruth, an attractive, energetic 35-year-old, earned $1,500 a year running a beauty shop in what used to be the Stephenson's dining room and Charlie made about $500 a year playing string bass with a dance band made up mostly of his neighbors. In more prosperous years, that money went for little "extras"—a sewing machine for Ruth, or a redecorating of the children's bedrooms. But in 1971 the money went for living expenses.

The Stephensons canceled an Arizona vacation they had planned for 1971 and Charlie said he couldn't afford to replace his 1949 Studebaker truck, even though he didn't "know how long that old clunker is going to last." Ruth wanted to remodel her cramped kitchen, where the family could barely squeeze around the table, but "that will just have to wait," she said.

Still, Charlie says he has known lean years before and he counts them a small price to pay for a life he loves. The Stephensons live in a gray, two-story wooden farmhouse about 30 miles south of Champaign, Ill., in some of the richest corn-growing country in the nation. Charlie's great-grandfather bought the land in 1896, and Charlie has wanted to farm ever since he was a boy. "I don't care if I ever leave the country. I'm happy riding my tractor and watching the corn grow," he says.

The Stephensons began renting the farm in 1965 when Charlie's parents retired and moved to a nearby town. Charlie built an addition to the red barn to house his machinery and converted a little white shed, where he and his father raised bees, to a repair shop. Ruth grows a few vegetables in the backyard, but they don't amount to much, and an old canning cellar next to the house is unused.

Most mornings the family is up at 6:30, with a little prodding from Ruth, who fixes a large breakfast of bacon and eggs. Bob and Sue go to school in Atwood, eight miles away. By 8:30, when the bus pulls into their driveway to pick them up, Charlie can be found in a pair of yellow rubber coveralls, cleaning and polishing his equipment near the barn.

"He can't sit still. He always finds something to fix," Ruth says. Dressed in jeans and a sweater, she makes a fresh pot of coffee for her beauty parlor customers, who often stay to sit around the kitchen table and chat. "Charlie wasn't happy when I set up shop, but we needed the money. I'd make more in town, but then I wouldn't be here when Charlie and the kids need me," she says.

After dinner, Sue, 12, washes the dishes while Bob, 14, feeds the dog and five cats. Sometimes Ruth drives the kids to a 4-H meeting or a basketball game, but most often she sews or reads while the children do their homework or watch television. Charlie, whose collection of musical instruments clutters Ruth's sewing room, often shuts himself in the kitchen with his string bass or tuba and plays along to whatever jazz music he can tune in on the radio.

Charlie and Ruth both say they cherish the independence of farm life and the solitude of the flat, peaceful countryside. Cities terrify them—"I never know my way around and the traffic ties my stomach in knots," Charlie says—and they rarely journey farther than the small towns nearby to shop or see a movie. On weekends, Bob and Sue tear over the near-deserted roads in the area on a minibike and go-cart, and they canoe on a river close by. Charlie and Ruth say they can't imagine their children growing up amid the congestion of city life. (Bob and Sue, however, can imagine it. Both speak of feeling "kind of left out" living in the country.)

Before the family moved to the farm, Charlie worked in a nearby chemical plant. He earned a steady $7,000 a year, but Ruth hated his working nights and Charlie hated the routine of the job. "I never felt important. Any farmer who complains ought to punch a time clock. I'd much rather work for myself."

But the price tag for working for yourself is high. Farming requires plenty of capital, and young farmers starting up have a tough time finding financing. While an urban family might mortgage a house or buy a car on credit, most farm families also must borrow heavily every year to pay for new equipment and for the cost of running the farm.

Charlie built up a $36,000 debt within a year after he took over the farm. Besides the Stephenson homestead, he rents another parcel of land nearby. He started out with a combine and his father's dilapidated plow and tractor. But the machinery broke down, and each year since Charlie has had to borrow more money for newer and bigger equipment. He was $25,147 in debt in 1971 and his interest payments amounted to about $1,200. "It just about breaks me keeping up with my machinery," he says ruefully.

That debt makes the Stephenson farm a touch-and-go operation. "One really bad growing season" would mean bankruptcy, Charlie says. There have been some close calls. The family barely scraped through the winter of 1966 after poor weather had damaged a crop and slashed their income. Two years later, hot weather scorched the corn, leaving some cobs with only a dozen kernels.

Despite all that, Charlie clings to the hope that hard work will overcome the travails of unpredictable weather and fluctuating prices. Even after the low prices of 1971, he said that he expected to get his "feet on the ground" any year now and that he was convinced that someday Ruth could stop working and the family would have the time and money to do things it couldn't afford then.

But it seems unlikely the future holds much of

either leisure time or money for the Stephensons. The two of them can just barely manage the farm during the months of planting and harvesting. To cut corners, Charlie haunts farm sales during the winter, looking for bargains on used equipment. He also spends hours in his shop tinkering with his machinery to save money on repairs. What's more, the only way Charlie can add income to meet rising costs is by farming more land. He calculates that he and Ruth could stretch their time and equipment over an added 100 acres.

But land around here is scarce. "If a fellow dies, his neighbors are bidding on his place before he has time to cool off," Charlie says. That's because most farmers are convinced greater production is the only way to make more money—despite the fact that a bigger national harvest often means lower prices. "The farmer is an eternal optimist," says one banker who finances farmers. "He always thinks prices will go up."

The Stephensons didn't participate in the government farm support program, which supported grain prices in return for limiting farmers' acreage. Charlie believes farmers would do better without "interference from Washington." Although he and Ruth are registered Republicans, he says the "White House cares more about its rose gardens than grain farmers." That, he says, applies to both parties. "There's not a dime's worth of difference between politicians. I wouldn't trust one if he told me the sun rose in the east."

Charlie's main concerns in 1971 were closer to home. He wasn't able to make a payment on the storage bin he added, and that meant a bigger debt the following year. He worried about his retirement—but has never been able to start even a meager savings account. "I ought to set aside $1,000 a year, but the cash just isn't there," he says. Charlie carries "only enough life insurance to bury myself," and concern about what would happen to his family if he lost his health gnaws at him constantly. He says an accident or a prolonged illness would ruin him.

But he counts his good friends on neighboring farms as insurance against that possibility. The sense of community survival is strong here, and Charlie often helps his friends out during planting or harvesting season. In return, a neighbor butchers a steer for the Stephensons, and the family, as Ruth puts it, "eats steak at hamburger prices all winter." Charlie likes to recall the time a friend broke his leg just before harvest time and his neighbors banded together to bring in his crop. "What city folks have friends like that?" he asks. "Money can't buy them."

Good times are shared, too. Nearly every Saturday Charlie plays with the dance band in some nearby town. Ruth spends the afternoon getting dolled up and then feeds the kids while Charlie dons a green jacket, black trousers, white shirt, bow tie and cummerbund. The band members' wives sit together at the dance and afterward the whole group drives to one of the farms for their own party.

The Stephensons count as vacations the three or four weekends a year that Charlie plays a dance 60 or so miles away. The whole family comes along and stays in a motel, usually chosen for the size of its swimming pool. Bob and Sue often swim until 1 a.m., to Ruth's dismay, and the family has dinner at a restaurant. "They should have a taste of city life," says Ruth, an Arkansas girl who admits she is still uncomfortable when confronted with more than one fork.

On Sundays, the Stephensons drive to church 15 miles away. They go to the 8:30 a.m. sermon together, then the children attend Sunday school while Charlie and Ruth attend Bible class. By the time she has cooked dinner, Ruth is usually drowsy enough to fall asleep sitting in an armchair; she and Charlie usually rest most of the afternoon. Bob takes off with a shotgun, trailed by Sue, who isn't allowed to hunt, and tries to flush a pheasant from the brush in the ditches along the road.

Charlie says he expects that someday Bob will take over the farm from him. By that time, he hopes, it will be a more thriving and less debt-laden enterprise. Meantime, he doesn't let hard times get him down. "I wouldn't trade places with anybody," he says. "I never expected to get rich."

—SUSAN B. MILLER

The New Pioneers—1971

When Sue and Eliot Coleman sit down to eat in their tiny one-room house, they use tree stumps instead of chairs. When they need drinking water, Sue walks a quarter of a mile through the woods to a freshwater brook and hauls back two big containers hanging from a yoke over her shoulders. And when the Colemans want to read at night, they light kerosene lanterns.

The young couple—Sue is 26, Eliot 31—aren't the forgotten victims of rural poverty or some natural disaster. They live as they do out of choice. They have deliberately given up such luxuries as indoor plumbing, store-bought furniture and everything that electricity makes possible. They have no telephone, no automatic mixer, no TV set.

With their two-year-old daughter, Melissa, Sue and Eliot are trying to escape America's consumer economy and live in the wilderness much as the country's pioneers did. They grow about 80% of their own food and spend only about $2,000 a year on things they can't make themselves.

The Colemans have been living this way two and a half years, and they're proud of their accomplishment. "If you listen to Madison Avenue, we don't exist," says Eliot. "They say it's impossible to live on $2,000."

The Colemans are among a tiny but apparently growing number of young couples, often from middle-class families, who are taking up the pioneering life, or "homesteading" as it's often called—though today's pioneers usually can't get free land from the government as early homesteaders did. Favorite homesteading areas are New England, the Pacific Northwest, the Ozarks and Canada. Sue and Eliot have 40 acres of thick forest 30 miles south of Bucksport, a small town near the central Maine coast.

No one knows just how many people are taking up homesteading. The Colemans say they personally know about a dozen couples. A neighbor of the Colemans, Helen Nearing, 67, who with her husband, Scott, now 87, retreated to a homestead in Vermont in the early 1930s and later moved to Maine, says "a lot of people, more than 100, are getting land and living off of it."

There's no doubt that interest is growing. In 1954 the Nearings wrote a book on the subject called "Living the Good Life." Only 10,000 copies were sold in the 16 years up to September 1970. But nearly 50,000 were sold in the following 10 months in a new edition.

People are turning to the pioneer life for a variety of reasons. Many are inspired by the philosophy of the Nearings, who lived 20 years in Vermont before they found ski resorts and other signs of modern civilization crowding in on them. In their book, the Nearings said they originally retreated to the land to find "simplicity, freedom from anxiety or tension, an opportunity to be useful and to live harmoniously." They arranged their lives so that, after working to produce what they needed to live, they had ample time for "avocational pursuits" like reading, writing, hiking and simply talking.

Some modern-day homesteaders have political motivations. "I don't want to earn a lot of money because I don't want to pay taxes to a government that's been lying about Vietnam and its intentions of solving social problems," asserts David Wilson, 27, an architect who is homesteading with his wife and two children in Maine. His wife Debbie, 28, agrees. "We're just totally exasperated politically," she says.

Others homestead because of interest in ecology and organic farming. "They're interested in life styles that will let them live well while doing good things for the earth," says John Shuttleworth, editor of Mother Earth News. The magazine, a year and a half old, has already built a circulation of 60,000 with advice on buying land, building pioneer-type homes and organic farming.

A chance to be alone with one's family also attracts some. "We've been invited into communes, but we aren't interested at all," says Mr. Wilson. "We have a tremendous need for solitude and privacy."

Communes are shunned by many homesteaders, in fact, on the ground that they tend to attract hangers-on, drug users and other undesirables who aren't really prepared to cope with the rigors of homesteading.

For Sue and Eliot Coleman, a desire to escape the consumer economy, a chance for real independence and a deep interest in organic farming all played roles in the decision to homestead. But their backgrounds would hardly indicate that they would someday try to live like pioneers.

Eliot, a short, solidly built man with blue eyes and a full head of unkempt, prematurely graying hair, is the son of a Manhattan stockbroker. He graduated from Williams College and worked on Wall Street as a broker trainee himself for a short stint. He soon gave this up to go to Middlebury College in Vermont, where he won a Master's Degree, then wound up teaching Spanish at Franconia College in New Hampshire. There he met Sue, who was a student. A pretty young woman with soft features and shoulder-length brown hair, Sue is the daughter of a vice president of a suburban Boston bank.

After marrying, the two came to their decision to homestead largely because of the inspiration of the Nearings' book, "Living the Good Life."

"We stumbled across the book while looking for yogurt in an old general store" in New Hampshire, Eliot says.

Sue and Eliot became vegetarians, as the book advocated, and spent $2,000 of their $5,000 in savings to buy their land in Maine. During their first two months in Maine in the fall of 1968, the couple virtually lived outdoors, their only shelter being a cramped three-foot-wide homemade camper body in which they slept. By day, Eliot chopped down trees and removed stumps until he had a clearing large enough to build a house on.

Using books and manuals as guides, the Colemans constructed an 18-by-22-foot cabin, using cedar posts for a foundation and planks of rough-cut wood he bought from a lumberyard for the floor and walls. Eliot's tools: a hammer, saw, level and carpenter's square. Total cost: about $700 for the materials.

"We would have built it out of logs, but it was October and we thought it would be good to get out of the camper" with winter approaching, says Eliot, almost apologetically. Logs would have had to be cut and would have taken longer.

The next spring was marked by the birth of Melissa, by natural childbirth and in the home—but with a doctor in attendance. It went without a hitch.

Since building the house Eliot has concentrated on clearing more land of trees, using axes and other hand tools, and has so far cleared four acres. A half-acre has been planted with vegetables and fruit trees.

Today it's hard for a visitor from the city to imagine that the Colemans' house and garden and orchard were once part of the thick forest of fir and spruce that surrounds them. The homestead, named "The Greenwood Farm" and set about a quarter of a mile off a dirt road, is striking with its carefully arranged rows of vegetable plants and small apple and pear trees that make up the half-acre front yard.

The Colemans grow 35 varieties of vegetables, including parsnips, asparagus, spinach, kale and lettuce, as well as strawberries and cantaloupe. A few rows of plants are covered with sheets of thin glass held together by wires and known as "cloches." They are, in effect, portable greenhouses. The Colemans also have a small greenhouse built into the front of their house. All this allows them to get a jump on the short Maine growing season, about 140 days, by planting vegetables while snow is still on the ground.

"By the beginning of March, we were eating radishes and lettuce," says Eliot proudly.

Watching the Colemans at work on a typical day provides some insight into just how they have accomplished as much as they have. Eliot and Sue arise at 5:45 a.m. After dressing in his customary brown corduroy pants, green short-sleeved work shirt, red pullover sweater and brown rubber boots (because of the moist Maine weather that keeps the ground damp and muddy), Eliot heads out to a one-acre field behind the house to remove tree stumps.

Eliot chopped the trees down a year ago, but the 150 or so stumps must still be cleared before the field can be plowed for planting. With a full overhead swing he chops at the largest roots of a stump's base with an ax and then switches to a pick and hoe to further loosen the stump. Finally, wearing gray work gloves, he wraps his hands around the foot-thick stump and pulls it out with a heave.

As the sun breaks through the early morning fog that hangs over the trees and raises the temperature about 10 degrees to nearly 70, Eliot soon finds himself soaked in sweat and removes his sweater. By the time he's ready to come in for breakfast at 7:30 a.m., he has already removed four stumps. He hopes to be able to plant a quarter of an acre of the field in corn this summer.

Inside the house, Sue, dressed in baggy brown work pants and a red and white striped pullover blouse, is also busy. She has already started the wood-burning stove, using paper and some small twigs to get it going, and now she's grinding wheat into flour using a cast-iron hand grinder. Later she will use the flour to make chappaties, an unleavened bread that resembles the Mexican tortilla in appearance.

Does she miss any of the modern kitchen conveniences most women her age long for? Not at all, she says. "I just thoroughly enjoy doing things by hand," she says. "Like grinding wheat. I'd much rather grind it by hand than use an electric grinder or blender."

Besides, Sue contends, her kitchen has its own versions of many modern conveniences. For instance, she can regulate the heat on her stove and oven according to the type of wood she uses. For moderate temperatures she uses softer wood like spruce or birch and for high temperatures she uses apple or cherry wood.

The house, although it consists of only one room, is divided into four areas—the kitchen with the wood-burning stove and two counters with storage shelves above and below; a dining area with a picnic-style table and the wood stumps for chairs around it; a living room area with two benches, built into the wall and covered with thin red mats, that serve as couches and as the lids of storage areas; and a sleeping area consisting of a large double bed built into the wall about five feet off the floor to take advantage of the rising heat in winter. Two-year-old Melissa sleeps in a corner of the bed. The only obvious signs of contemporary life are the books, many of them on organic farming, that fill the bookshelves built into one wall. There is also an old pedal-operated sewing machine. The toilet is an outhouse about 50 feet from the house.

After starting the stove and grinding the wheat,

Sue heads out back to a small fenced corral to milk one of the Colemans' three goats (the other two, daughters of the oldest goat, are too young to give milk as yet). The first month or so after Sue started milking the goat, which must be milked twice a day, her arms and hands were sore, she says. "But they say you develop milker's hands after a while," she observes.

The milk is mostly for Melissa and is stored under the house in a narrow cellar that serves as a refrigerator with a temperature between 37 and 47 degrees. The cellar also is used to store root vegetables such as potatoes, carrots, beets and turnips, which are kept in shallow boxes of sand to retain freshness and which are eaten throughout the winter.

Breakfast is extremely simple—apples dipped in ground sesame seeds, which look like a gray paste but have a sweet, candy-like taste, and ground oats with raisins and goat's milk. Until a year ago, the Colemans had about a dozen chickens, which they used for eggs, but they gave the chickens away when they decided they didn't care all that much for eggs.

After eating, Eliot pulls out a black three-ring notebook in which he records such things as the daily weather, the date certain crops start growing and how well they grow. He also charts the chores that remain to be done.

"I think we ought to start the parsnips now," he tells Sue. "Last year I think we started a little late. This way they should have better roots," he says. Sue agrees, and he makes a notation in the notebook.

The rest of the morning Eliot spends pulling stumps out of the ground and Sue divides her time between making chappaties and pulling weeds in the garden. Melissa occupies herself playing with pots and pans or wood sticks or simply wandering around the garden, chattering contentedly.

Lunch, served at about noon, consists of potato and onion soup and fresh chappaties. Dessert is chappaties spread with peanut butter and honey, both store-bought.

After lunch Sue walks through the woods, with Melissa following, to their three-foot-wide brook. There she fills two containers with three gallons of water each, enough to last for two days, and carries them on her shoulder-yoke back to the house.

For washing and bathing water, they use a well near the garden (it hasn't been tested for pollution as the spring water has, so they don't use it for drinking). They heat the water on the stove. An oval three-foot-long metal tub serves for both bathing and clothes-washing. For soap, the Colemans use store-bought Ivory bars and flakes.

The afternoon finds Eliot turning over a patch of the garden to prepare it for the planting of parsnips and carrots. An important part of the preparation involves mixing compost, or fertilizer, into the soil. The compost is a mixture of seaweed the Colemans get at the nearby seashore and horse manure and leftover hay from a local horse farm as well as remnants from their meals, all of which has decomposed for months. "If I had to buy all sorts of chemicals and fertilizers as most farmers think they have to do, I really would be in a cost-price squeeze," says Eliot.

The Colemans have made some concessions to 20th century technology, however. They have kept a small Jeep, a Volkswagen truck and a portable Zenith AM-FM shortwave radio, all of which they owned prior to moving to Maine. Eliot likes to listen to the news and weather reports once or twice a day, and on Sundays he tunes in classical music.

The Jeep and truck trouble the Colemans, however, as symbols of modern technology and sources of pollution. Eliot says they're considering selling the truck and hitching a trailer to the Jeep whenever they need to haul things. "At the rate we're going, we'll have an ox and mules in a few years," he says. "Who knows, then if I want to go to town, we'll hitch up an oxcart and make a day trip out of it."

Otherwise, the Colemans have been able to divest themselves of nearly every sign of middle-class life. They gave two electric blenders and other appliances they had received as wedding gifts to friends before leaving Franconia College. Eliot discontinued his Blue Cross and Blue Shield coverage as well. "Health insurance is served on the table every meal," he says, expressing the belief of many organic food devotees that food grown without artificial fertilizers and made without chemical additives improves health.

Eliot admits to some misgivings about forsaking insurance. "I had that fear every suburbanite has, but living like this, you get over it," he says. "I figure if anything happens, I'll find a way to cope with it." If he should face a sudden big doctor or hospital bill, he figures, he will pay it off over a period of years.

Eliot and Sue still retain some ties to the money economy. During the spring and summer Eliot does gardening and other odd jobs for local residents three or four mornings a week, for which he is paid $2 to $2.50 an hour. Sue also has done some part-time secretarial work. Together, they were able to earn about $1,400 last year. They earned another $350 from the sale of surplus vegetables from their garden—mostly peas and lettuce —to neighbors and tourists, for a total income of $1,750. The remaining $250 they spent came from the last of the savings they had when they moved to Maine.

Of their $2,000 in expenses, the largest single item —about $750—went for gasoline, repairs and registration of the truck and Jeep, which is another reason Eliot wants to get rid of one or both vehicles. Another $500 went for food they couldn't produce themselves, such as raisins, vegetable oil, nuts and 100-pound bags of wheat, oats and rice, which together last about a year.

About $200 went for new gardening and construction tools, and another $200 went for household items like kerosene for lamps ($14 for a year's supply), windowpanes and soap. Other purchases included books, seeds, food for their three goats and dental bills. Clothing expenses are minimal; they are still wearing clothes from their prehomesteading days, and Sue sews what else is needed.

This year Eliot hopes to take another step toward self-sufficiency by selling $800 worth of vegetables and fruit. That will mean he and Sue will still have to earn about $1,200 to meet their $2,000 budget. Eliot regrets having to take on odd jobs, however, because "the time I put in doing that I lose here (working on the farm)," he says.

He is especially wary of both himself and Sue being gone at the same time. One day in 1970 when both of them were away, one of the goats got out of the small corral behind the house and ate a whole patch of lettuce. "That set us back a month-and-a-half," says Eliot. "It was a real disaster."

After Sue has washed the lunch dishes, swept out the house and taken a short nap with Melissa, she joins Eliot in the garden and helps plant the seeds. To Eliot, the time he spends in the garden is probably the most fulfilling part of any day. "It's a beautiful feeling when I'm out here with a hoe and I think that this is something man has been doing for 4,000 years," he says as he turns up clumps of earth. "We could have the TV and refrigerator if we busted our tails and planted every square inch of our 40 acres, but that wouldn't be any fun."

That's not to say, however, that the Colemans don't have some expansionary plans. Besides clearing more ground for farming, they want eventually to build a larger house and turn the small one into a workshop. Sue is a potter and Eliot a skilled woodworker. They haven't had the time or the facilities to practice their crafts since moving here. But they believe that once the farm is in the shape they want it, they—like the Nearings—will have ample time for "avocational pursuits."

At about 5 p.m. Sue goes in to begin preparing dinner, and by 6 p.m. Eliot's 12-hour day has ended and he comes in to wash from a large bowl of hot water. Then he flicks on the radio to catch the weather forecast for the next day and sits down to a bowl of rosehip soup (the family's main source of vitamin C), which is made from the fruit of the rose plant. Next comes a tossed salad of lettuce, kale, grated beets, carrots and chopped onions, all grown in their garden. The main course is oatmeal topped with natural sesame oil and steamed kale. Dessert is apples.

Following dinner, Sue and Eliot relax by reading and playing with Melissa. The kerosene lamps add to the relaxed mood by giving a soft glow that just allows for reading.

Perhaps every month or so Sue and Eliot get together socially with a young engineer and his wife who live nearby and have a child Melissa's age. Occasionally they visit friends in Bangor or see the Nearings. That's about the extent of their social life, however. They haven't been to a movie in about three years, and they say they don't ever feel the need to go to a restaurant to eat, preferring their own organically grown food.

Hard as their day-to-day work may seem, the Colemans appear to find it a small enough price for the satisfactions of their life. "I'm working 16 hours a day for survival," Eliot says. "This isn't any game I'm playing.

If I don't grow enough, it's that much less to eat this winter." But at the same time, he says, "We find, every day, we're just so happy here."

During the winter, things slow down a lot, says Sue. She spends her time mostly knitting, sewing, cross-country skiing and reading. Eliot chops trees when the snow isn't too high and joins her in reading and skiing. "In the summer, you're rushing around trying to grow your food," Sue says. "Winters are rest times around here."

—David Gumpert

The New Pioneers—1973

When Eliot and Sue Coleman bought 40 acres of land 30 miles south of Bucksport, Maine, in the fall of 1968, their aim was to escape America's consumer economy and subsist as much as possible off the land. Their domain at that time consisted of thick forest, and day after day they furiously worked to carve an independent existence out of the woods.

By the summer of 1971, Eliot, who was 31 years old at the time, and Sue, then 26, had cleared a small area of their heavily forested acreage and had set up housekeeping with their two-year-old daughter, Melissa, in a tiny one-room house. They were growing about 80% of their own food and were spending about $2,000 a year on necessities that they couldn't produce themselves. A small part of this spending money came from their savings; the remainder came from Eliot's odd jobs, from Sue's part-time secretarial work and from their roadside vegetable stand.

The Colemans' ultimate goal then was self-sufficiency. "I'm working 16 hours a day for survival," Eliot said at the time. "This isn't any game I'm playing. If I don't grow enough, it's that much less to eat this winter."

How have Eliot and Sue Coleman fared since then? Today, the Colemans are on the verge of that hard-sought self-sufficiency. "We're almost over the hump," Eliot says. "The idea was that the first five years would see the farm supporting us. I think we'll do it."

A few numbers underlie Eliot's optimism: a herd of six goats, versus half that number two years ago; a two-room house versus one room then; and nearly two-

and-a-half acres under cultivation versus only half an acre in 1971. Possibly most important, income from operations— vegetable and fruit sales—totaled $2,400 last year, up from only $350 two years earlier.

"When you look around," Sue Coleman says, "it's so satisfying."

Indeed, when you look around the Coleman property, it is apparent that they have been exceedingly busy during the past two years. Not only is their house larger—a change in part brought about by the birth of a second daughter, Heidi, last New Year's Day—and their acreage expanded, but there are also new buildings: an eating and milking shed for the goats; a combination woodshed-workshop; a roofed open-air stand for selling surplus vegetables.

All of these changes, within the context of the Colemans' existence, are vast, but they haven't been accomplished without attendant headaches and sacrifices— and one of these sacrifices has been abandoning from time to time the homesteader's aim of shunning modern technology. Eliot, for example, finally decided that pulling all the tree stumps out of his land by hand was too time-consuming, and last summer he hired the owner of a back hoe (a construction-type machine) to pull out the stumps at a cost of $25.

"It took me the better part of a summer to destump half an acre," Eliot explains. "He did a quarter acre in two hours . . . it was like the jolly green giant had come in to help me make the garden."

But then with a bigger garden, tending the land began to demand more time than the Colemans could provide. The upshot: This summer, they are using the services of a 21-year-old Boston youth, who, having expressed an interest in organic farming, is working for the couple as an unpaid volunteer. (The youth lives in a small "quasi cabin" on the Colemans' land.)

The Colemans are glad to have the additional help. They do worry, however, about setting an important precedent. "I don't want to turn it (the farm) into somebody else's work," Sue says. In fact, so fast have they grown, Eliot says, that "pretty soon it's going to be g-r-o-a-n."

Their growth has also meant borrowing some of the marketing practices from the consumer economy that they say they are fleeing. To attract area residents and tourists, they have put up signs advertising their produce at strategic locations in nearby towns. They also take out small advertisements in two area newspapers. For the sake of good customer relations, Sue has mimeographed about a dozen different recipes built around vegetables that the Colemans sell. And as a "premium," last summer they passed out fresh flowers with each vegetable sale.

Sue and Eliot Coleman have discovered that their dramatic increase in vegetable sales has brought them the headaches that go along with any small business. Eliot, for example, says that he now must spend considerably more time with customers than was necessary in earlier years. During the summer of 1970, he recalls, "someone would come in and buy something, and then I'd go back and hoe for an hour." But last summer, he says, "it got to the point where I got nothing done" in the gardens because he was so busy servicing customers.

The additional visitors have also posed some problems for the Colemans' coveted privacy. Last summer, Sue says, people sometimes came past the vegetable gardens in front of the house and peeked in the windows of her home. "They wanted to see how the freaks live, I suppose," she says. This summer, to forestall such invasions, the Colemans have put up signs announcing that the house is "private."

Yet in many respects, the Colemans' life is unchanged from two years ago. They still have no electricity, indoor plumbing or telephone. When they want light, they use kerosene lanterns. When they need drinking water, they obtain it from a brook a quarter of a mile through the woods away from the house. When they cook, they use a wood-burning stove. And in place of a refrigerator, they use two cool cellars for food storage.

Sue and Eliot are also still vegetarians, an eating philosophy that they adopted (for humanitarian and nutritional reasons) when they first set up housekeeping in Maine. They also still work long hours, still use tree stumps instead of chairs and continue to shun modern entertainment—they haven't seen a movie in five years—and modern conveniences.

But even this Spartan existence requires money—more money, in fact, than Sue and Eliot have thus far been able to earn from their crops. The woodshed-workshop, for example, cost $100 in materials, the addition to the house $300. Another $300 went for glass covers to protect certain vegetables in the expanded gardens. And as a consequence of these and other costs, largely related to expansion, the family's living expenses last year rose to $3,000 from $2,000 in 1970.

In order to make ends meet, the Colemans borrowed $1,000 at the beginning of last year. They used $400 of this money to ease their 1971 deficit and $600 to supplement 1972's $2,400 in vegetable and fruit sales. In addition, Eliot took a temporary carpentry job last January and February that netted him $600, enough to tide the family over until crop sales began again this year.

Eliot hopes that the carpentry work will be his last outside job. This summer, he and Sue expect to earn $3,000 from crop sales, allowing them to meet their expenses without additional borrowing and thereby fulfill their dream of self-sufficiency. But their big problem now, they say, is to decide whether to try increasing their sales even further.

"If we want," Eliot says, "we could sell 10 times as much as we do now. We could make our gardens bigger, earn more money and pay people to come in and do things for us. But I'd rather do it myself." He adds: "There comes a point where you make so much money you can't do what you want to do."

If Eliot Coleman makes homesteading sound easy, it should be noted that he has had his share of setbacks. This spring, for example, he discovered that only 20 of the 50 small raspberry plants he had purchased the year before were alive, the dead plants having succumbed to neglect. ("We bit off more than we could chew" in the spring of 1972, he explains.) On top of this, he noticed one day that 125 cauliflower sprouts had disappeared, apparently eaten by deer or rabbits; at 65 cents a sprout, this represented a total setback of about $80—the couple's biggest loss ever.

But such reverses haven't dampened the Colemans' enthusiasm for their way of life. "If anything," Eliot says, "I'm more of a freak for this life and the land than I was before." (His enthusiasm has apparently infected the couple's eldest daughter, Melissa, who now is a cute four-year-old. The child contemptuously refers to people who eat fish and meat as "other people" and to cities as "bad.")

Despite their qualms concerning expansion, the Colemans won't be standing still in the future. For example, Eliot plans to clear an additional three acres of land for a goat pasture and grain crops. He also envisions the construction of a second greenhouse.

But all that probably won't come until at least after this summer and maybe not until next year, Eliot says, principally because of all the strains of the past two years' expansion. A new house that would have allowed their existing cabin to be converted into a pottery workshop for Sue has been put off at least five years, he says.

"I feel in a way I've blown it here and I've let the place get too big," Eliot reflects. "I sometimes think that maybe I'd like to pick up in 10 years and go someplace else and be even more self sufficient."

—DAVID GUMPERT

One Family's Farm

Like his great-great-granduncle and namesake a century before him, William Penn Tuttle III returned to the family farm only after a fling at life in the big city. He drove a taxi, sold used cars and was a salesman for Campbell Soup Co. in Boston for a year after graduating from Tufts University with a degree in sociology.

"I had never spent time in the city, so it seemed exciting and challenging," says Bill, a 27-year-old bachelor. "But it turned out to be just a bunch of people I couldn't stand." Nor could he stand the city's long theater and restaurant lines and the traffic jams. "It got so I hated getting up in the morning to get where I was going," he says.

So he went home, to the Tuttle family's rolling vegetable farm near Dover, N.H., on either side of State Highway 16. "I was dumbfounded, absolutely dumbfounded when he walked in that spring morning," says 53-year-old Hugh Clarke Tuttle, who for years had been heartsick over his son's disinterest in the farm. "I thought he had a tiff with his boss, and for a year I didn't believe he'd stay."

Bill's homecoming means that the Tuttle farm—which dates back to 1632 and is apparently the oldest farm in the U.S. still owned and operated by the same family—will continue at least through the 11th generation.

It also signals a new era for the farm. Whereas the entry into commercial farming assured the farm's survival into the 20th century, Bill is busy building it for the 21st. He's expanding the number of products sold in its retail store, called Tuttle's Red Barn. And he's using sophisticated marketing, advertising and cost-cutting techniques in an effort to increase sales and profit. The farm itself is going through a transition, from sole proprietorship to family corporation.

While its 245 acres are less than the 385-acre national average for farms, sales have more than doubled to more than $200,000 expected in 1974 from $72,000 two years before. That puts the Tuttle farm in the top 10% of U.S. farms—those with annual sales of more then $40,000 that produce 61% of the nation's food, according to the U.S. Department of Agriculture. (Of the nation's 2.8 million farms, only one million are "commercially productive"; at least one million are small, part-time enterprises with under $2,500 in annual sales, the department says.)

Moreover, the Tuttle farm in 1974 received one of 14 annual Ford Motor Co. "Ford Farm Efficiency Awards." "Hugh Tuttle has an overall high-quality, efficient, well-run operation," says Cliff Ganschow, editor of the "Ford Almanac," an independent publication that makes the selections. "He's representative of a really top food producer, but he's got a perspective that makes him special."

That perspective is rooted deep in the farm's history, which dates back almost to the beginning of Colonial times. Its history has closely paralleled that of the nation's farming industry. "For centuries, farms were self-sufficient," says Wayne Rasmussen, Agriculture Department historian. "Horse-drawn farming came in during the Civil War, signaling the beginning of commercial farming, or producing for a market. Machine farming came in after World War II, 1954 being the first year that tractors outstripped horses.

"Those changes prompted the death of most 17th- and 18th-century farms, as they were forced to merge to justify the efficient use of improved methods." Since the 1935 high of 6.8 million farms, 20,000 to 30,000 have folded each year.

That the Tuttle farm survived is a testament to its Quaker tradition, its sense of timing, and its people. The Tuttle story, pieced together from historical records, old albums and family memories, began in Bristol, England. King Charles I granted an apprentice barrel maker—John Tuttle, or "Immigrant John," as he's now called—the right to about 20 acres on "Hilton's Point," a nine-year-old settlement between the Cochero and Bellamy Rivers. John emigrated on the ship Angel Gabriel, and like most early settlers, he lived as a subsistence farmer.

Life was difficult. "The new land had everything going for it after crowded 17th-century England, but only the tough survived," Hugh says. "The climate was poor, the primeval forest thick, the topsoil only two to three inches."

Three centuries later, to learn just how difficult, Hugh himself cleared a virgin patch of forest, treated it with lime and manure—and plowed crops under for four years. "I started planting buckwheat, which grows anywhere. But it still took six years before the topsoil built up to the minimum needed. I don't know how my ancestors did it," he says.

By the second or third generation, the Tuttles had become Quakers. That was a key factor in the farm's survival because by Quaker tradition the youngest son inherited the farm, along with the responsibility of feeding old parents, grandparents and maiden aunts. (Older sons got $100 and were told to go seek their fortune.) "The patriarch demanded ironfisted obedience. Thus, as soon as the youngest was old enough, he damn well knew he'd get the farm," Hugh says.

So stern were they that when Joseph, of the sixth generation, found his daughter with a farmhand, he sentenced her for life to a back bedroom with "passing and repassing privileges." Family members couldn't talk to her; meals were sent up on a tray.

"Captain John," son of the immigrant, first prevented the farm's breakup. An almost legendary figure, he held such town posts as judge, clerk, selectman, treasurer, and captain of the military company in charge of fighting Indians. When his son died at age 26, he demanded that the farm go to his grandson, rather than to the offspring of the man his son's widow had remarried.

Very little is known about the 18th-century Tuttles. Town records were destroyed at least three times by fire. And early graves were unmarked for fear that Indians would see how depleted were the settlers' numbers or that they would rob treasures and scalps. It's known that Eliah, of the fourth generation, replaced the old log cabin in the late 1700s with a fine white house, occupied today by Hugh's father. It's suspected that the Tuttles were either rebels or pacifists (in line with Quaker tradition) during the American Revolution.

In the 19th century, Joseph's eldest son, William Penn Tuttle (the current Bill's namesake and now referred to as "Uncle William"), was an eccentric innovator, or a "man ahead of his time," as the Tuttles now say. He left the farm at 21, spent 30 years in the sawmill business in Milton, N.H., and spent whatever money he made. After the death of his younger brother in 1874, he returned to the farm, anxious to test some far-fetched ideas.

He built the first greenhouses in the county, possibly introduced the horse-drawn plow, and began selling a few surplus items to passersby. Just to prove he could do it, he grew crops considered exotic at the time—grapes, horseradish, potatoes and cranberries. "People came for miles to see them," Hugh says. "He had to dump the grapes, but he picked them just to see how many he could grow."

Such ventures were hardly profitable. By the time he died in 1911, Uncle William had expended three wives' dowries, driven the farm into disrepair and nearly outlived his younger brother's son, to whom he wanted the farm returned. But in 1909 when his 18-year-old grandnephew, William Penn Tuttle II (called "Penn"), approached him with the idea of selling the farm's produce to independent grocers, Uncle William spotted the novelty and chuckled. "Thee go ahead and see what thee can do," he said.

That first season Penn did $700 of wholesale business, ushering in the farm's new commercial era. A consummate businessman, he kept accurate records and had one of the first telephones in the area installed to sell products. He hauled them personally at first by wagon and later in one of the first farm trucks in the area. He believed strongly in standing behind only the finest products. "The secret is honest salesmanship," he wrote in a family album. "I charge fancy prices, but customers gladly pay for my service."

As his business expanded to surrounding towns, Penn expanded the farm, adding about 100 acres, a new barn, a new house and two greenhouses. By 1930 he was growing 25 vegetable varieties on 20 tillable acres.

By the 1950s, however, the Tuttle farm was being hurt by the spread of supermarkets, which were unwilling to pay top dollar for superior produce. But rather than fold, it began in 1956 to sell directly to consumers. "It was a desperation move," says Hugh, who by then had taken over from his father. "We had to jump someway, and a roadside stand seemed logical because we had built up a clientele."

To make it work, Hugh hired an experienced manager to advise him on price markups, display of products and the handling of customers. He bought an adjoining 39-acre farm, adding 10 tillable acres to his land. His wife, Joan, tended the store. And they managed to eke out a modest living. Sales from the stand totaled $500,000 between 1956 and 1969.

That it survived is a tribute to Hugh, Penn's second son. (His eldest died in an accident; his youngest is a prep-school administrator. The Quaker tradition of the youngest son's inheriting the farm broke with Penn, who was the older of two sons.) From the beginning, Hugh wanted to be a farmer. "A greatuncle asked me when I was five where I'd be going to school. I said I wouldn't be going, that I would become a farmer," he recalls.

Hugh's mother, an educated woman from Brookline, Mass., encouraged his older brother to attend Harvard. Hugh followed, but left after three years, in 1943, to return to the farm. Tall, well-built and ruddy, Hugh is a man in love with the soil. "My soul is at ease—I have no trouble sleeping, I'm home with my family, I commute only 100 yards. I die a little in the fall when my plants die, but I'm reborn each spring."

In some ways, he's downright old-fashioned. He introduced irrigation only after he was forced to draw water from town hydrants during a disastrous drought in 1953. (The farm now has four ponds.) "Until then we thought irrigation was merely drought insurance. Now we see it as a production tool," he says.

Likewise, he didn't bring in tractors until he expanded the farm in the 1950s. "My father and I both loved working with horses," he says. "We felt that tractors posed serious problems in New England—they go so fast it's hard to spot rocks, for one."

(Hugh isn't averse to some change. He experiments with plant varieties, having added Danish lettuce, seedless cucumbers and other items to his line, and he donated his $2,000 Ford award to the state university for greenhouse vegetable research. He's also a community activist, having spent years in state agricultural organizations and in local and state politics. He's a bank trustee and director of the largest utility in the state).

Hugh couldn't have been more overjoyed by his son's homecoming. For years he had forgone capital improvements, convinced that the farm would finally die. "I couldn't see a small New Hampshire farm bucking the trend, nor did I want to pass on a dead horse to my son," he says.

Today, Tuttle's farm is anything but dead. Its 60 tillable acres include 20 acres of corn, five acres of beans, four acres of peas, two acres of strawberries, 2½ acres of tomatoes, 1½ acres of lettuce, and smaller quantities of 69 other vegetables. Fourteen field hands work six months a year, planting every week from April to July, harvesting from May through October. Nine salespeople man the Red Barn six days a week, setting out just-washed produce in wooden bins, distinguishing Tuttle products with signs that say "Our Own." More than 500 customers stop by the store each day; on peak weekends as many as 1,500 come.

While incorporating the old Tuttle philosophy of selling only the finest produce at higher prices, Bill has expanded it in the past two years. Aiming to build a complete produce center, he began buying supplementary fruits and vegetables directly from the New England Produce Center near Boston, to which he personally drives 60 miles three mornings a week to start shopping at 5 a.m. These items accounted for 17% of last year's $122,000 gross and made up an even bigger percentage in 1974.

Father and son are expanding in other ways: They've added a cooler and two greenhouses and have expanded the stand and parking lot. In 1975, they planned to begin planting on five rented acres. Tuttle's advertising budget has jumped to $4,000 for thrice-weekly ads in local papers, up from once every two weeks. Merchandising techniques now include putting corn near the back so that customers have to pass by other vegetables, giving away samples of fried eggplant and stocking collard greens to attract Southerners from a nearby Air Force base. On tap for the future may be an expanded house-plant business, a "pick-your-own-strawberries" operation, and a line of homemade baked goods.

Such expansion spells death for certain farm operations, such as its small herd of Angus cattle. "I always thought a farm should have some animals to look like a farm, but they aren't economically feasible anymore," Hugh says. The family also abhors the Red Barn's three checkout counters and cash registers, which cut down the store's former personal-touch ambience. "It's becoming a super tourist attraction," mourns Hugh's 22-year-old daughter Becky. Over Bill's protest, she persuaded her father not to buy corn when Tuttle's isn't available. "It's terrible—and would downgrade ours," she says.

Expanded operations also mean more problems. Increasingly, Hugh has had to contend with more and more government regulations. One proposed federal rule, defeated last year after farm groups lobbied against it, would have required that he post a sign each time he sprayed his fields with pesticides. "That would have meant 696 signs and, by town law, a permit for each one," he shudders. A new law that requires a minimum wage for farm workers will force up prices, Hugh says.

Yet the farm is going forward. While it comfortably feeds one family now—Hugh's annual profit is about $15,000 to $20,000—he hopes it will soon feed two and possibly three future families: Bill's and Becky's, as Hugh feels she'll settle here, and maybe even Lucy's. His

29-year-old daughter came home in August 1974 to take agricultural courses at the University of New Hampshire after teaching in Paris for seven years. He has no intention of dividing the farm among his children. "This farm must be operated as a whole to survive economically," Hugh says. "If it were split, it would die."

—LIZ ROMAN GALLESE

Boss Men

It was a long, hard summer in 1969 for the vegetable growers of Northampton County, what with low prices, a drought that hurt early strawberries and cabbages and, finally, 17 consecutive days of thundershowers. To cut their losses, some farmers let whole fields of waterlogged potatoes and tomatoes go unpicked.

The farmers aren't the only victims, however. Abandoned crops also mean hard times for the Negro migrant workers who pick crops every summer on Virginia's flat, sandy Eastern Shore. Some found only two or three days work a week that summer. Others worked more regularly but made less money than usual because muddy fields made vegetables hard to pick.

"I can't recall a year as bad as this one," said Leon James, a Negro who had journeyed up from Florida for the harvest for two decades. While disappointed, Mr. James didn't seem to be particularly worried, and with some reason. He is neither farmer nor worker.

Mr. James is a migrant worker crew leader, officially called a "farm labor contractor." Vegetable farmers in Virginia, Florida and New Jersey pay crew leaders such as Mr. James for recruiting, transporting, paying and supervising the Negro workers who pick or grade the crops. In good years and bad, farmers need migrant workers—supplied by the labor contractors. Startlingly enough, in good years a crew leader can make up to $50,000.

Mr. James said he was lucky to clear $200 a week in Virginia in the summer of 1969. But, back home in Fort Lauderdale, Mr. James and his brother Louis have "a few investments" which, he said with a thin smile, "ought to see us through." They include three small apartment buildings, a fleet of 16 taxicabs and a bus service that delivers domestics to and from work. The seed money for these businesses came from the migrant-worker enterprise.

The crew leader is a singular figure in American agriculture. To many a farmer he is indispensable, as the primary—or only—source of labor at harvest time. To the worker he employs, he is an absolute authority. In the labor camp, the crew leader is foreman, paymaster, Dutch uncle, money lender, grocer, policeman, judge and jury. Police and other local authorities leave the operation of the shabby, isolated labor camps almost entirely to the crew leaders.

Wielding that kind of power, crew leaders live very well indeed. Most are former field hands themselves. At best, the crew leader seems to be a benevolent despot, using his control of work, money, housing and transport for the benefit of his workers as well as himself. As crew leaders go, Mr. James is favorably regarded.

But Harvey Robuck, a thin, weathered Negro in another crew, was unhappy about his crew leader. "He picked me up in Bonita (Florida), where I was drunk at the time," says Mr. Robuck. "When we got to this place, he told me I owed him for driving me up here and for my keep. Then I had to borrow some dollars so I could eat and buy a little wine. Then comes Saturday, and he says I only got $4 coming. Five days I worked, and I only got $4 coming. I went out and had me a good time, and Monday I got to borrow a few more dollars."

Mr. Robuck found himself trapped in the migrant debt cycle. Only the crew leader can find him work or give him a ride back to Florida. The crew leader also had him constantly in debt, furnishing wine and cigarets on credit ($1.50 for an 89-cent bottle of port and 50 cents for a 35-cent pack of cigarets). The crew leader's two assistants warned Mr. Robuck that he would be "cut bad" (with a knife) if he tried to leave the camp while he owed money.

Mr. James says he doesn't operate that way. His workers in 1969 were getting $1.50 an hour, average for potato graders. And he shuns the petty usury and wine sales used by some crew leaders to inflate their incomes. Nor does he charge for lodgings and transportation to the fields, as some leaders do.

His workers also enjoy relative comfort and sanitation. They live in a red cinder-block compound put up by Eastville farmer Robert Hoeffner at a cost of $30,000 in 1968. Mr. Hoeffner employs the James crew each year. The compound replaced an old, dilapidated labor camp. The Spartan 10-foot-by-12-foot rooms are no cozier than Army barracks, but they are clean, dry and free from bugs and rats. Mr. James insists on periodic cleanups and on keeping the compound neat. When a worker broke down a door during a dispute with his wife, Mr. James made him pay the $38 repair bill. Some camps have dismal accommodations; one crew leader has housed his charges in an abandoned potato shed.

The James crew benefits also from the help of outside agencies in the Eastville area. Day-care centers tend the workers' infants, and each day a yellow bus hauls children to a special school for migrant families in nearby Accomack County. There is some dispute about the school's quality, but it keeps the children out of the fields and feeds them solid, balanced meals to supplement a diet that often consists only of pork, beans, cabbage, rice and bread.

The migrant health project in Nassawadox, 10 miles north of Eastville, sends Dianne Diaz, a young nursing student, to visit the camp every few days to check for sick persons who should visit the project's mobile clinic. "Mr. James is awfully cooperative," she says. "The first time I went there, he went through the camp and shouted, 'We got a nurse outside. Any of you need help, get on out there.' I've had no trouble at all since then."

Mr. James' relative benevolence contrasts with the approach of Ozell Suggs, who has a 47-man potato-grading crew. Mr. Suggs feels he "doesn't need to worry much" about health and sanitation. "If they want to live like animals, I let them," he says. His formula for success: "You keep them happy. You get them work and you pay them. Some of them like a little wine, so you give them some money and let them go get it."

How many of his crew members owe him $20 or more for wine? "I guess pretty much most of them." Does he always get his money back? "I get my money back. I don't have no trouble I can't take care of." Mr. Suggs is six feet, two inches tall, and weighs 285 pounds. The Wellington boots he wears make him look even bigger.

Crew leader Ray Gordon also has an offhand attitude toward the welfare of his workers. The migrant health project determined that one of them had tuberculosis and arranged admission to a sanatorium in Charlottesville. Mr. Gordon balked at letting the man —"my best worker"—go for treatment during the harvest, even though the worker's positive sputum test indicated he was a hazard to the other laborers. Only when the county sheriff intervened was the man taken for treatment.

Leon James spends much time recruiting workers, traveling to "the places where migrants hang out" in Norfolk, Baltimore and Philadelphia. But the supply is short. In desperation, he hired 10 local Negroes to help out the crew in 1969.

Ironically, hard-bitten crew leaders like Mr. Suggs seem to have less trouble keeping workers than "humane" ones like Mr. James, who reports a high turn-over. From May 1 to Sept. 15, 1969, he employed more than 100 different workers in his 35-to-40-man crew. "To keep migrants today, you've virtually got to keep them prisoner," says one veteran Eastern Shore farmer.

This old hand might have added that he, and the other farmers relying on migrant labor, are in something like economic bondage to the crew leaders themselves. The five trucks that haul potatoes from Mr. Hoeffner's fields belong to Mr. James. For every 100 pounds of potatoes unloaded from the trucks, Mr. James gets a flat fee, which doesn't change, regardless of the price that Mr. Hoeffner gets for the produce on the market.

In the summer of 1969 Mr. James showed up at the Hoeffner farm a week late and a few workers short. That cost Mr. James and his crew of potato graders some money, but it hurt Mr. Hoeffner and Raymond Briggs, the other local farmer using Mr. James' crew, far more. Most of their potatoes had to be shipped in bulk by the truckload, fetching a lower price than if there had been time to grade out the better-quality potatoes in 50 or 100 pound bags.

Mr. Hoeffner acknowledges that Mr. James "has his faults," but, he says: "I don't think I could find a better crew leader." That was a reason for the new housing that Mr. James' workers got, and it also accounts for Mr. James' fee, which is higher than is paid to some other crew leaders.

Mr. James and his brother Louis started the crew in 1946 when they respectively were 17 and 19. By 1969, the business included the five potato trucks and a green and black school bus used to transport the workers. Louis has retired from crew leading to manage the other family enterprises in Florida. Leon was accompanied by his father Allen, a garrulous, 250-pound former field worker who handled many of the dealings with the laborers.

"He cusses at you a lot," says one field hand, "but he don't mean it."

Both father and son live modestly. They stay in the compounds with their crew (Allen is separated from his wife, and Leon leaves his wife and 11-year-old son in Florida during the harvest), drive old-model cars and usually dress in rumpled green work uniforms. "They're kind of like grandpappies to all of us," says crew member Roosevelt Buchanan, to whom the Jameses lent $283 for an auto repair bill before coming north to Virginia.

Not all crew leaders live modestly. One evening, Roscoe Simmons, a Negro crew leader, was observed in a chauffeur-driven, air-conditioned white Cadillac, roaring past a farmer in a battered pickup truck on a local highway.

However, crew leaders, flamboyant or modest, find themselves in a dwindling industry. Migrant labor is fading from the Eastern Shore. In the 1950s, there were 14,000 migrant workers in Northampton and Accomack counties each year; by 1969 there were about 6,000.

The families that have grown vegetables on the Eastern Shore since the days of slavery find that the sandy soil and unpredictable weather won't support crops large and uniform enough to compete with produce grown by mechanized methods and trucked in from other regions. Many large farmers and farming corporations were switching to rye or soybeans, which don't need migrant labor. Some smaller farmers were quitting the business altogether and using their land for more profitable enterprises, such as tourist camping grounds.

Adding to the crew leaders' problems are the unscrupulous "free lance" crew leaders who ignore federal and state regulations governing migrant labor. These contractors arrive without a work contract, house their crews wherever they can find a roof and seek work by undercutting the usual prices.

Leon James estimates that migrant laborers—and crew leaders with them—will have vanished from the Eastern Shore by 1980. But the Jameses will keep at it until the end. Leon says simply, "I like crew leading," and his father Allen says, "I do it to keep busy. It's all I got to do now."

—JAMES MACGREGOR

Slim Pickings

The migrant farm worker tends to dream of the "nice field" or the "$100 deal" ($100 a week) that lies ahead in a rosier future. "Just wait until we get to Ohio," says an aging Mexican-American laborer. "The tomatoes will be as big as pumpkins, and we'll have work every day."

Listening to such talk is dispiriting, for the plentiful crops, good living conditions and generous pay never seem to materialize. The old worker concedes that he got to the Michigan fields too late last year to do well. "But this year—you wait and see," he says smilingly.

I saw. In a week-long stint as a cucumber picker in 1969, I found that the $100 wage was more likely to be $40 to $60 a week, with grinding labor, dreary housing, inadequate food, and an economic system that fixes the worker in bondage to his crew chief. The hopeful chatter in the evenings comes to seem like idle dreaming indeed.

Nobody knows precisely how many migratory farm workers there are in the U.S.; the Senate subcommittee on migratory labor estimated that in 1969 there were 276,000, most of them originating in Florida, Texas and southern California.

The migrants at my camp came mostly from three distinct groups: Mexican-American families that make a yearly circuit from Florida through Michigan, Indiana and Ohio back to Florida; white Americans who head north from Tennessee and Kentucky in their dilapidated autos each summer in search of temporary work; and rootless single men who roam the countryside working long enough at each stop to buy food and drink. At my camp they ranged in age from five to more than 70.

Around Keeler, Mich., the crops are cherries, apples, cucumbers and an array of other fruits and vegetables. Picking cucumbers, I found that you have to scramble to make $1 an hour.

On the first morning, a burly Mexican-American overseer thrust a plastic half-bushel basket into my hands and informed me that I would be paid 45 cents for each time I filled it. "Get to work," he snapped, repeating the command in Spanish to make sure I understood.

It turned out that the field in question already had been picked once. The big cucumbers were gone. Left on the vines were the smaller vegetables, and over-ripe ones. A cheerful 13-year-old girl showed me how to beat the system, putting the less desirable vegetables in the bottom of the basket and the acceptable ones on top.

"That will fool them at the truck," she said smilingly. It did. My baskets were accepted without comment and emptied into a big crate. Even so, I only managed to deliver six baskets in the first three hours of work, for earnings of 90 cents an hour. Occupational hazards included scratches from the prickly vines, cramps in the lower back and eventual aching of the thigh muscles. Other workers said I was proving quite productive, however.

Jesus Flores, the crew leader who had recruited and transported most of the workers to Keeler to work as a team, said that he only got paid 10 cents for each basket of cucumbers delivered by his laborers. "It isn't as easy for me as you think," he said.

Mr. Flores was a better than average crew chief. For instance, he didn't charge for transportation to the field each day, as some chiefs do. But he had extra sources of income. On payday, he deducted 20 cents from my wages, "for the Government." But he never asked for my Social Security number. And I had difficulty collecting from him for my last day's work—90 cents for two baskets of cucumbers. Finally he pulled a $1 bill from his pocket and handed it to me. "Keep the change," he said.

Home for me in this labor camp was a nine-by-12-foot cabin, one of 10 such structures standing back to back in a double row. Fire had blackened part of the walls and roof, but my cabin had two rusty metal beds, a decaying couch, and an old refrigerator and stove— amenities lacking in many camps.

However, it lacked plumbing, shelves and closets. Old paper bags and newspapers had been used in an unsuccessful attempt to plug cracks in the walls and ceiling. It was stifling hot in the daytime, freezing cold at night. On the walls were mementos of earlier residents —drawings of cars, mountains and houses, and the penciled message "Ignacio loves Maria."

One of my roommates, a veteran on the migrant-labor circuit, had had the foresight to bring a faded blanket with him. This was Jess, a 24-year-old Mexican-American whose right arm and leg are withered from a childhood disease. Jess had worked in the fields since quitting school in the sixth grade.

My other roommate and I made do without blankets or other bedding. He was Steve, a 19-year-old Tennessean who recently spent 37 days in jail for driving a stolen auto. Now a drifter, he has been trying to save enough money to head for Detroit, Chicago, or back to Tennessee. "Anywhere," he said, "just to get out of here."

The cabin stank of sweat, grease and rotting fruit. Fully clothed for warmth, I slept on a torn and fetid mattress above springs fastened to the bed with bits of rope, old electrical cord and coat hangers. Flies and mosquitoes circulated freely.

It is an existence that a young single man can endure. For families, it is grim. The camp's four crude privies stand in a nearby apple orchard. They smell awful, and many workers simply wander into the orchard. Water, for bathing, drinking and cooking, comes from two taps at one end of the row of cabins.

Next door a family of nine was crowded into a cabin the same size as ours. Ignacio, eight years old, told of the routine in the cabin: "We just turn off the light to keep mosquitoes away and go to bed. My mother then tells us stories about herself and our family. Then we pray together for a while and go to sleep."

Beans and pork chops were my diet for the week I was at the camp, although my roommates and I occasionally slipped into nearby fields in the early morning or late at night and snitched a few peaches, apples or potatoes. Jess and I ate from a single plastic plate using the only spoon in the cabin to scoop the food onto pieces of bread. We shared an old water jar.

For medical problems there was a clinic in Keeler, two miles from the camp. But crew leaders weren't eager to have workers take time off for the excursion. Next door a five-year-old boy had received 11 stitches for a cut in his foot, but hadn't returned to the clinic for removal of the stitches. The foot was infected and swollen. "I was supposed to take him back to have the stitches out" said his mother. "I've mentioned it to Jesus (the crew leader) a couple of times, and he's changed the subject."

So resigned are migrant workers to their situation that they often fail to seek out existing facilities. Church groups and state agencies offer various kinds of assistance that the migrants seem not to know about. For instance, two miles from the camp the Michigan

Migrant Ministry had an office that provided canned food and clothes to migrant families at little or no cost. But most migrants were ignorant of this service.

One reason is that some growers and crew leaders make it difficult for outside agencies to establish communications with the workers, viewing such visits as a possible challenge to their authority. Most of the camps in this region are surrounded by signs reading "Keep Out" or "No Trespassing." One local farmer beat up an employe of the United Migrant Opportunity Inc., a Federal Office of Economic Opportunity agency. The agency vowed to move against attempts to isolate workers from outside help.

The fatalistic outlook of the workers is evident in the words of a Puerto Rican woman, a mother of six: "There just isn't enough work around here, and the picks are too scarce when we do work. When I was little, my mother told me that we were poor, and no matter how hard I tried, I would always be poor. You know what? She was right. That's why I don't kill myself working. It isn't worth it."

Why don't the workers unionize? "I don't mess around with that," said one Mexican-American man. "Look, the minute you start talking of unions and strikes you'll get kicked out of the camp. And how the hell are you going to feed yourself and your family if you don't have any money, no work and no place to go? It's bad enough the way it is, but it could be worse."

The camp where I worked was owned by Ferris Pierson. "I'll be the first to admit that some of this housing isn't the best in the world," Mr. Pierson said. But he believed his cabins were above average. "At least we provide them with bottled gas, stoves and refrigerators," he remarked.

Mr. Pierson said he even allowed some cucumber pickers working for another grower to stay in his cabins. "They literally came to me with tears in their eyes," he said adding, "What are you going to do with these people? What I want to know is what the Government is going to do with these people—what anybody is going to do with them."

He saw hard times ahead for growers, who are faced with rising costs and diminishing profits. "The farmers are just about at the end of their rope," Mr. Pierson claimed. And, he said, the crew leaders are apprehensive too. "I think a lot of them are more worried about what's going to happen years from now than about the conditions today."

—RAUL RAMIREZ

Struggle in the Fields

In the dusty little town of Coachella, Calif., in the heart of grape-growing country, breakfast sometimes isn't a very healthy way to start the day.

One morning in the late spring of 1973, I and the Rev. John Bank, a 33-year-old priest and official of the United Farm Workers (UFW) union, sat eating breakfast at a local restaurant when a dozen burly members of the International Brotherhood of Teamsters ambled in. The atmosphere, at first jovial, became more hostile as the Teamsters and the slender priest traded jibes. Suddenly, a husky Teamster approached Father Bank, hit him in the face and broke his nose.

The incident dramatizes the explosive dispute between the Teamsters and Cesar Chavez's UFW over representation of the state's 250,000 farm workers. Before it is over, the struggle may decide the fate of Mr. Chavez's union. Thus, the outcome is likely to have wide-ranging impact on the chaotic farm labor situation in California—the top crop-producing state—and in the nation as a whole.

High-level efforts to reach an accommodation were being pressed by both sides in 1973. Teamsters President Frank Fitzsimmons has conferred with AFL-CIO President George Meany, who is backing the Chavez union. But some insiders question how effectively a truce between the two labor leaders could be enforced in the vineyards, where emotions won't necessarily be tempered by soothing words from Washington.

After his stunning success in 1970 organizing farm workers, many of whom are migratory, Mr. Chavez has seen his power base crumble while the Teamsters—and, some say, the growers—have seized the upper hand. The defections began in the early spring of 1973, when growers in the Coachella Valley (130 miles east of Los Angeles) spurned Mr. Chavez at contract renewal time and signed instead with the Teamsters. Within two months, Mr. Chavez estimated the UFW lost about 150 of 190 contracts in the state to the Teamsters, and the UFW's dues-paying membership shrank to 10,000 from 40,000 in the same period.

Never one to take a defeat lying down, Mr. Chavez called for a strike and a new nationwide grape boycott. His strategy was to apply enough economic pressure on valley growers at the height of the grape harvest so that they renounce the Teamster contracts and sign once again with the UFW. That, the UFW hoped, would prompt the state's big table-grape growers in Delano, about 125 miles north of Los Angeles, to renew their UFW contracts, which expired in late July 1973. Table-grape contracts were the strength of Mr. Chavez's movement.

It didn't work out that way, and the Teamsters remained aggressive competitors of the UFW. But Mr. Chavez vowed to fight on. Even an anti-Chavez grower admits, "Mr. Chavez will be a viable force in the labor movement as long as he chooses to be."

The 1973 strike wasn't very successful. UFW pickets—not all of them farm workers—numbered anywhere from 800 to 1,100, but growers said they had no difficulty recruiting labor or shipping crops on schedule.

To counteract the picketing, the Teamsters paid beefy "guards" $67.50 a day to "protect the workers in the field," according to a spokesman. It was one of these guards who hit Father Bank.

The story of how the UFW's 1970 grape victories eventually turned sour is riddled with angry rhetoric and charges and countercharges on all sides. But the roots of the dispute go back many years and stem from the exclusion of farm workers from the original National Labor Relations Act, which provides collective bargaining procedures for industry. Although growers, Teamsters and the UFW say they favor some legal protection for farm workers, the three groups differ about the form such law should take. The American Farm Bureau Federation, a large grower's organization, wants to ban secondary boycotts and harvest-time strikes—both favorite Chavez tactics. The Teamsters favor farmworker coverage under the National Labor Relations Act, a plan Mr. Chavez opposes because the act also bans secondary boycotts.

The UFW claims the current struggle stems from a long-standing "conspiracy" on the part of big growers and the Teamsters, the nation's biggest union, to destroy the small farm workers union. Mr. Chavez considers the contracts the Teamsters signed with Salinas lettuce growers shortly after the UFW 1970 grape triumphs were announced as a "stab in the back." As evidence of collusion, the UFW officials are fond of pointing to an affidavit filed by an official of a large commercial farming concern charging that Salinas lettuce growers actively sought Teamsters representation for their workers—without asking the workers.

The UFW is also unhappy with signs that the Teamsters and the American Farm Bureau Federation seem to have reconciled their once-sizable differences. Teamsters President Fitzsimmons was invited to address the Farm Bureau's convention in December 1972, and he called for an alliance of labor and agriculture "when that alliance works for the mutual benefit of the farm worker and his employers." He closed the speech by telling the growers, "If you believe you have a grievance, let's sit down and talk it over."

The conspiracy theory is endorsed by the AFL-CIO, with whom the UFW is affiliated. Mr. Meany's strong support of the Chavez union has placed the prestige of the 13.5 million-member labor federation squarely up against the 2.4 million-member Teamsters, who were tossed out of the AFL-CIO in 1957 on corruption charges. Mr. Meany has termed Teamster activity among farm workers "union busting" and has given the UFW $1.6 million to help sustain the strike.

The Teamsters deny charges of conspiracy and contend they started organizing in Coachella Valley only after the workers themselves petitioned for the big union's representation. The UFW, however, maintains that some of the reported 4,200 names on the petitions were obtained by "fraud, coercion or deceit."

Moreover, the Teamsters claim they have long been interested in organizing farm workers, because field strikes directly affect the jobs of drivers and cannery workers, whom they already represent. Also, they say, mechanization is shrinking processing-plant jobs and forcing workers into the fields. "The more organized we are," says a spokesman, "the better we can service our members from an economic standpoint." The UFW argues that until 1970, when the Chavez movement made its gains, the Teamsters had only one contract covering field workers.

The growers also deny there is a conspiracy, and as evidence, point out that the Teamsters pact will drive their costs up about 25% the first year. They blame UFW administrative blunders and Mr. Chavez himself for losing the grape contracts. "I think he blew it," says David Smith, an attorney for growers of half the valley's grape crop. Pointing out that Mr. Chavez had all the major table-grape growers in the state on two- or three-year contracts, he adds, "If his workers had been happy and his people had managed their union in a businesslike way, there's no way in the world another union could have taken the workers away from him." And, while praising Mr. Chavez for his "inspirational leadership," Lionel Steinberg, one of two Coachella Valley growers to renew with the UFW, complains the labor leader is a "very poor administrator and an inflexible negotiator." In Mr. Steinberg's view, local UFW officials are often "more interested in social revolution and agrarian reform than in assuming the normal role of union business agent negotiating between management and labor."

But according to the UFW, the real reason the growers oppose the union is its "hiring halls"—a concept that changes the traditional agricultural power relationships. Under the hiring hall system, workers are dispatched to growers on a seniority basis, rather than through foremen who, before the halls were set up, often extracted fees from both employers and workers.

Growers contend it is actually the workers who object to hiring halls, because the system splits up family working units by sending some family members to one farm and others to a different farm. The UFW concedes this happens, but emphasizes the importance of seniority rights so that jobs can be stabilized. "Without seniority, there isn't a union," Mr. Chavez says.

Moreover, the UFW claims that some foremen, loyal to growers, sabotage the hiring-hall procedure in order to embarrass the union. One favorite ploy, Mr. Chavez claims, works like this: A grower asks his foremen to place a request at the hiring hall for, say, 200 workers. In fact, the foremen request only 150. When the 151st worker comes to the hall for a job, the union tells him all jobs are filled. Then the worker goes to the foremen, who tell him jobs are still available.

The workers themselves haven't had a voice in determining who represents them. Mr. Chavez has called for a secret ballot election and predicts that 90% of the workers would vote for the UFW. And many of the state's retailers, who are major targets of the UFW's boycotts, have placed full-page ads in California newspapers calling for a secret ballot.

The Teamsters, however, aren't anxious for elections. As one official says, "When we believe the workers are supporting us, we see no reason for getting involved in a long, drawn-out procedure for elections."

—WILLIAM WONG

Harvest Nomads

For a man the government classfies as a "migrant farm laborer" Frank Wood fares pretty well.

Occasionally, for instance, he flies his private twin-engine Beechcraft to Las Vegas, where he "sees a few shows, shoots a little dice and, you know, just relaxes." If the dice roll the wrong way, it probably won't be a disaster: Mr. Wood's income averages $45,000 a year.

The burly, 40-year-old Oklahoman earns the high life as a "custom cutter," one of the nation's most nomadic jobs. With eight big red Massey-Ferguson combines, 11 heavy trucks and the private plane, Mr. Wood and his team of 18 husky workers follow the American harvest, cutting and hauling grain crops for a fee.

Custom cutters are a rare breed. There are only about 400 in the whole country, and most people don't even know they exist. But cutters perform a major task on the farm and their job of bringing in the harvest is big business. Mr. Wood's team alone cut 32,000 acres in 1970—at a clip of one acre every 10 or 15 minutes. "Farmers know we can cut crops faster, cheaper and better than they ever could," the soft-spoken Mr. Wood says.

The highly mechanized operation, which uses $250,000 worth of equipment, promises to cut crops exactly when they're ready. Cutting even a day or so too early or too late can sharply affect the price of a crop, and serious timing errors sometimes make the grain completely unsalable.

A look at Frank Wood's modern, high-speed operation vividly shows how life on the farm has changed.

Mr. Wood recalls his teen-age days on his father's Oklahoma farm. Perched in the open on the hard steel seats of the combines of those days, you "chewed a lot of bugs, ate a lot of dust and got a sore rump," he says.

In contrast, the youthful combine operators working for Mr. Wood nowadays ride in glass-enclosed, air-conditioned cabs. Although Mr. Wood turned down optional built-in radios in the cabs, he did allow the young drivers to do their own interior decoration. The result: Fancy imitation leather seats, pinups, portable stereo tape players and even carpets on the floor. "As long as these kids do their job, I could care less what they put in those cabs," says Mr. Wood.

But even with carpeted combines, cutting can be one of the most grueling jobs going. It appeals only to the restless. Most big-time cutters like Mr. Wood roam nine months a year searching for ripening crops. They typically start in Arizona and Texas when the first of the season's wheat crops mature in late April. During the spring and summer months, they push as far north as the Montana and North Dakota wheat fields. Later in the year many turn back south, cutting Arkansas and Mississippi soybean fields until December.

Until Mr. Wood married in 1951, he often slept underneath his combines. Then he bought a camper for his family and a converted bus for the crew to use as a dormitory. In 1957 he decided that staying in motels would be less trouble.

Thus a typical day harvesting wheat began at "Bair's Little 5 Motel," in Bozeman, Mont.—with a 5:30 a.m. call. During a breakfast of fried eggs and country sausage, Mr. Wood quickly outlined the day's work as his men sleepily nodded assent.

Although the men can't begin cutting crops until the morning sun burns the dew off the fields, they often switch on the combine headlights and work late at night. When rain threatens to spoil a ripe crop, they reach a frantic pace. The men more than once this year have cut grain for 24- and even 36-hour stretches. (This is possible only if weather conditions reduce the dew problem.)

The pace leaves little time for leisurely meals. For lunch, a hungry worker frequently drives his combine past the boss's truck, grabbing a cheeseburger and milkshake from Mr. Wood's hand. In contrast, harvest hands before World War II could commonly count on a relaxed two-hour noonday break with lunch at the farmer's table.

To make good on his guarantee to cut crops at precisely the right moment, Mr. Wood keeps close tab on weather conditions, scrutinizes local grain reports and surveys planned cutting sites from the air. When the crop appears to be ripe and ready, he moves in for the crucial test, plopping a sample handful of grain into his mouth and testing the kernels. If the grains taste "mushy," they aren't ripe enough. But if they prove just resilient enough to the bite, the crew of shaggy-haired young men and wizened farmhands get the order: "Move 'em in!"

Mr. Wood readily admits his odyssey of more than 4,000 miles would be impossible without the modern combine, which has vastly changed harvesting.

First used on a wide scale in the late 1930s, the combine cuts grain, separates the kernels from their husks and cleans them. It replaces the reaper and the threshing machine. The latter, which separated grain kernels from their husks, required a 12-man crew. Besides eliminating this crew and cutting labor costs, introduction of the combine speeded the harvest.

Mr. Wood's eight combines cost $16,500 each in 1970—and they became steadily more expensive in each succeeding year. Mostly because breakdowns would slow his high-speed operation, he replaces them all each year. The six-ton 1970 models had 20-foot blades in front. These big rotating blades don't actually cut the grain; they merely direct it into the combine where smaller blades chop it. The 1970 models were 150% more productive than their four-ton ancestors of 25 years ago, the custom cutter says, and further efficiencies have been built into later models.

Another change in recent years: All Mr. Wood's combines and several of his trucks fly small, dime-store American flags. One of the combines has a faded "Love It or Leave It" sign. Mr. Wood, who often wears an American flag pin on his sweater, claims the flags were the crew's idea, not his—but he's clearly glad to see Old Glory flying.

The combine—and World War II food shortages—created the custom cutter's job. In 1942 the Government told farm implement makers they could continue to make combines instead of tanks, but buyers had to agree to harvest neighboring fields after finishing their own. Many farmers who accepted the deal found the extra work lucrative. After the war, they continued the practice; many made it a full-time job and called themselves "custom cutters."

Since the war the increased efficiency of new combine models has helped improve the pay of both boss and crew. Mr. Wood's cutters with several years experience received about $4 an hour in 1970. New men got half that, still about 50 cents more per hour than most farm laborers could expect. Every man is also paid half his motel bill (four or five men to a room) and gets all the hamburgers he can eat while on the job.

Rusty Jardee, who worked on the crew in the summer of 1970, says the job "is the best thing that ever happened to me." A sophomore that year at Montana

State University, he said the $1,500 he earned would "more than pay for my school costs."

But the grueling work and nomadic conditions don't appeal to everybody. Even Mr. Wood's son Gary admitted that after a year in college, "I've decided to do something different." This didn't surprise his father. Amused by the suggestion that some might find his nomadic life adventurous, he says there is little excitement to be gained from "living in every two-bit motel between Texas and Canada."

Accompanied every summer by his wife and three youngsters, the custom cutter also worries that his family may feel the strains of suitcase living "more than they let on." Yet Mrs. Wood says she wouldn't have it any other way. "I just go wherever Frank goes," she says. While on the road, she occupies herself reading magazines and collecting rocks. During the school year, she and two of the youngsters lived at home in Hobart, Okla. Gary was attending the University of Oklahoma.

But as he looks forward to that Las Vegas trip, Mr. Wood considers himself far better off than many others who rely on farming for their incomes. While many big farmers earn high profits, thousands of hard-pressed small operators have folded in recent years. "Look at the bright side," he says, thinking of these smaller farmers. "We're among the few people left who can still make money off the land."

—DANFORTH W. AUSTIN

Riding It Out

The odds are that Jim McAllen will survive an economic crunch in the cattle industry with the same gritty self-reliance that has enabled him and his ancestors to ranch the semitropical South Texas land since the late 18th century.

The McAllens have withstood attacks by bandits and rustlers and made it through droughts, depressions, hurricanes and epidemics of cattle diseases. Like other Western cattlemen, they are fiercely independent and they treasure their traditions proudly. Jim McAllen, who in 1974 at the age of 36 ran the 70,000-acre ranch with his 62-year-old father, has modernized operations considerably. But by breeding and inclination he remains a cowboy who knows that life on a cattle ranch still demands long hours of hard work, the instincts of a gambler and more than a little luck.

In 1973, all this resulted in record profits for cattlemen, but for much of 1974 Mr. McAllen and other ranchers were squeezed by rising costs and falling cattle prices. Costs of things that ranchers buy had risen 15% above a year earlier by mid-1974, and one expert estimated that keeping a cow or raising a calf for a year had risen to about $190 from $135 to $145 a year before.

Meanwhile, the prices Mr. McAllen was getting for the cattle he ships to feedlot operators for fattening had dropped. When price ceilings on live cattle were lifted in September 1973, he held back some calves for six months in hopes that prices would improve. Eventually, he sold them for $249 each, $10 less than the price allowed—and paid—under the price ceiling. In early June 1974, his calves (which weigh about 500 pounds each when he sells them) brought $140. "We're getting now what we got 10 years ago," he says.

The prices fell largely because of consumer reluctance to buy beef. Per capita consumption dropped to 110 pounds in 1973 from a record 116 pounds in 1972. Though organized boycotts seemed to go out of style, housewives continued to bypass the beef.

"We haven't seen this kind of reluctance this century," said David Stroud, president of the National Live Stock and Meat Board, a trade group, as prices dipped lower and unbought beef was jammed into freezer warehouses.

All sorts of dire warnings were being made as a result of this squeeze—that the cattle industry would go bankrupt, that production was being cut back and so less beef would be available in a couple of years, and so forth. As evidence, the doomsayers cited the fact that in May 1974, 10% fewer cattle were put on feed than in the year-earlier month. Even fewer would be fattened unless prices go up, they warned, urging that the government restrict meat imports.

But cow-calf operators such as Mr. McAllen are the key link in the production chain that moves a calf from the open range to the supper table. The feedlot operator, the meat packer, the purveyor and the retailer are all important, but the rancher determines the supply of cattle that is ultimately available to consumers.

Mr. McAllen says he can't afford to cut back his herds. "The land has got to produce," he explains. "If we were speculators or weekend ranchers, we could sell out. But when bankers, doctors and lawyers start folding up their investment ranches, it avalanches down on real ranchers." They apparently are responding much like Mr. McAllen: Even as ranchers' prices were falling, the Agriculture Department forecast another increase in the nation's beef herd for 1974—a 12% rise to 130 million head.

"The more we produce, the more we want to produce," Mr. McAllen says. "It's kind of bred into us. If the government would leave us alone, we'd produce what it wants. We don't want to be controlled by anybody, and we don't like agricultural products being used as a pawn in international deals." Predictably, Mr. McAllen is in favor of restrictions on imported beef. "We're the people that made the country—not the foreigners—and you should think of the home folks first."

Such bristling don't-tread-on-me attitudes are common among ranchers. "Cattlemen have always been able to weather the storm, but I don't think they've seen a storm like this—at least, not in the last two decades," a Department of Agriculture economist said in mid-1974.

But Mr. McAllen was confident that the ranch would come through the hard times as it has endured others for generations. "No other industry can suffer a one-third loss in a year and still operate," he says. He recognizes, though, that ranchers have image problems with both politicians and consumers.

"The housewife at the meat counter probably thinks the rancher made his money overnight and drives around in a Rolls Royce with steer horns on the hood," he says. "Actually, I'm a welder, mechanic, electrician, plumber, a helluva good windmill repairman—and last, I'm a cowboy."

He spends more time in his air-conditioned, radio-equipped pickup truck than in the saddle, and wields a welding torch as often as a lariat. But like the cowboys of a hundred years ago, his workdays stretch from dawn to dusk.

One recent morning, he arose as usual at six o'clock. By seven, he was swallowing a second cup of coffee and calling his father on the two-way radio in his office to ask if a certain section of fence would be repaired that day. Then he drove his 1974 truck through several pastures to check on windmills and water tanks.

East of a saltwater lake, he flushed two coyotes from their daylight lair. He grabbed a loaded rifle from the seat next to him and fired two shots, but the coyotes disappeared over a ridge into thick stands of mesquite. "The next shot's yours," he said with a grin to a visitor in the pickup cab.

The water-supply survey took until lunch time, which is to say until 1 p.m. Mr. McAllen's father, Argyle A. McAllen, refuses to set his watch to daylight saving time, so everybody on the McAllen ranch eats lunch an hour later than the hands on other ranches. ("The cows aren't on daylight saving, so why should we be?" Mr. McAllen's father says.) Mr. McAllen, his father and their Mexican-American hands share a lunch of meat and beans, served on tin plates stamped with the McAllen brand, a "rolling" S M (after Salome, Mr. McAllen's great-grandmother).

After lunch, he continued to drive to his outlying herds. Like most ranchers, he doesn't like to reveal exactly how many animals he has on the ranch. He does admit to having 4,000 cows, plus bulls and yearlings. The average calf crop runs 85% to 90% of the cows (about 3,800 calves) in the December-March calving season.

As he drove, the blue-eyed, sandy-haired rancher looked for the telltale signs of screwworm wounds, observed which cows are about to give birth and, once, hearded a bull from one pasture to another with the pickup.

The next morning Mr. McAllen drove to nearby Reynosa, Mexico, to buy cowboy gear for his ranch hands. While he and the proprietor haggled amiably in

Spanish over the price, Mr. McAllen examined each of the 10 pairs of chaps he wanted to buy. They settled on $56 a pair, compared with $33 a year ago. (Chaps of similar quality would cost about $75 in the U.S., he says.)

He winced as he figured the bill of sale. "He showed me his cost and asked if he could make $8 a pair," Mr. McAllen said. "He's got to make a living, too."

Mr. McAllen spent that afternoon behind shaded goggles welding a lock piece onto a gate. A sudden rain blew in from the south, cratering the red earth and driving several uncorraled horses beneath mesquite trees. As sparks sputtered from the welding torch, Mr. McAllen said: "My dad used to be a little leery about all the newfangled stuff I was doing. Then I built this barn we're standing in and he got interested pretty quick."

The ranch has about 130 horses that are used by the eight ranch hands to patrol the herds. Because their acreage is spread over three separated parcels of rolling rangeland, the McAllens usually load horses and riders into trailers and tow them miles from the ranch headquarters before deploying the cowboys among the cattle. Besides working the cattle, the cowboys have fences to mend, windmills to maintain, pens to build, brush to clear, foraging grass to plant, and wells and waterways to dig.

More than ranching goes on at the McAllen ranch. The first of now numerous natural-gas wells was drilled on the property in 1965. In the fall, wealthy deer and quail hunters (including such celebrities as Bing Crosby and Phil Harris) lease portions of the ranch and go hunting from comfortable, well-outfitted mobile homes.

Mr. McAllen and his pretty, blonde wife Frances regularly entertain friends and neighbors in their rambling, Spanish-style home (which is the main house of the ranch, located a few hundred feet from **Mr. Mc-Allen's** father's home). The single-story house, made from white clay blocks and topped with red tiles, is surrounded by a three-foot limestone fence to keep out rattlesnakes.

During one barbecue, Mr. McAllen confessed that "we buy most of our own beef." Holding up a thick, well-marbled steak, he adds: "We couldn't afford to raise beef like this." That's because it would take weeks of fattening on grain in a feed lot. He thinks, though, that more people will begin buying grass-fed rather than grain-fed beef because costs and prices would be lower all along the line from rancher to housewife. Grass-fed beef doesn't taste as good, he concedes, "but the consumer wants cheaper beef."

When he's not worrying about economics, Mr. McAllen devotes his time to professional and personal interests. He is on the boards of directors of the Texas and Southwestern Cattle Raisers Association, the Rio Grande Valley Ballet Foundation and the county historical society. He also paints, sculpts and has collected a wide variety of artifacts, ranging from arrowheads to antique guns. He is something of an inventor, too, having devised a multilock gate and a feeder that allows only bulls, not cows, to eat special feed.

Mr. McAllen grew up on the ranch, attended schools 35 miles away in Edinburg and won several Future Farmers of America cattle contests. After spending a year at Texas A & M University ("I didn't like school," he says), he leased a large ranch from his uncle and ran it for 11 years. In 1969 he entered a partnership with his father and two aunts, whose combined holdings now make up the McAllen ranch.

The ranching heritage of the McAllens is strong. Originally, the land was part of a Spanish grant in 1791 to a direct ancestor of Mr. McAllen. Several generations of McAllens and their relatives by blood and marriage have ranched it since then. (The city of McAllen, some 40 miles south of the ranch, is named after Jim's great-grandfather.) Many of the Mexican-American hands who have worked with the McAllens were born on the ranch and lived and died there.

Together, the owners and workers have defended the ranch against all sorts of calamities, including a 1915 raid by Pancho Villa's bandits. Bullet scars are still visible above a doorway and in the seat of a wooden chair in the elder Mr. McAllen's high-ceilinged home.

When he was 25, Jim McAllen married Frances, an Edinburg native who studied English at Southern Methodist University. "I thought sure I'd wind up in Dallas," she says with a laugh, "but Jimmy yelled at me to come back home, so I did." They have three daughters, who in mid-1974 were aged 10, nine and five, and a 15-month-old son, James Jr.

"I'd like to see James become a rancher," Mr. McAllen says, "but I won't insist on it. I'll leave it up to him."

Young James will grow up observing on his father's office wall a leather map of the ranch that bears a legend of the ranch's history and concludes: "That you our sons may carry on where we left off and fathom more."

—MIKE THARP

A Vanishing Breed

The early-morning sun is beginning to burn away the mist hanging in the draws and valleys as Monte Noelke reins in his horse. The sheep he has herded for an hour now mill restlessly behind him in a corral. He strikes a match on his leather chaps and cups the flame around a corncob pipe.

"I don't know," he says with sigh, leaning forward in the saddle. "I don't know whether the domestic sheep industry in the United States has to survive, but I know one thing—it's meeting its test right now."

Mr. Noelke is a sheep rancher outside the West Texas town of Mertzon and, according to some agricultural experts, he is one of a dying breed. While farmers and cattlemen have enjoyed recent boom years, the windfall has passed up most of the 162,000 sheepmen (some with flocks of 100 sheep or less) in the 50 states. Although rebounding wool and meat prices caused a minor resurgence in 1973, and though petroleum shortages threatened synthetic fabrics, droves of ranchers continue to leave the industry. More than 8,300 quit in 1972 as the number of sheep fell to its lowest level this century, 17.7 million. That was down 5% from 18.7 million in 1971 and was far below the record of 56.2 million sheep in 1942.

Sheepmen blame their plight on a flock of factors, ranging from weather to the fickle consumer. But the newest and most emotional problem—though probably not the most significant one—is predators. And when they say predators, they mean coyotes. Even environmentalists concede that coyotes kill some sheep, primarily lambs. "I think they ought to be compared to rats," one prominent Idaho rancher says.

In 1972, an executive order banned the use of poisons for predator control on public lands, which make up most of the sheep range in Western states. Ranchers contend that coyotes are driving them out of business. In Wyoming, for example, 117,800 sheep valued at $2.2 million were lost to predators in 1972, up from 112,000 sheep and $1.8 million a year earlier. So the ranchers are lobbying to bring the toxicants back.

(Environmentalists insist that misapplied poisons kill wildlife indiscriminately and that predators account for only a small part of livestock losses. They also fault sheepmen for poor range management and outmoded husbandry practices.)

Mr. Noelke admits that sheep ranchers have an image problem, which makes it tough for them to win many arguments, particularly political ones. "I think most Easterners picture us as some dumb guy in brogan shoes and a sheepskin coat who throws up fences to keep out the cattle," he says. Sheepmen have also been saddled with public outcries against eagle kills, against shooting of coyotes from airplanes and even for allowing rattlesnake hunts on their land.

They could probably live with their tarnished image, but they have also been losing money. Sheep pay off in two ways—wool and meat—and most recent years have been lean for both. The bottom dropped out of the wool market in 1971, largely because of the surge in popularity of polyester fabrics. In that year, average prices fetched a meager 17 cents a pound, 55 cents below the federal incentive payment—the amount guaranteed by the government to take up the slack in lean years. However, thanks to heavy Japanese purchases in the Australian wool market (which tends to govern world wool prices), the price in 1973 recovered to about $1 a pound.

And meat prices strengthened, too. But Americans' taste for mutton, always meager, continued to decline. Per-capita consumption of mutton in 1972 was a mere three pounds, down from almost five pounds a decade ago. Mutton consumption is far below the 115 pounds of beef the average American ate in 1972. Ironically, the states that produce the most sheep are among the most finicky when it comes to eating it. Most mutton and lamb are shipped to the East and the West Coasts and to Chicago for ethnic dishes.

Despite occasional respites, the downward trend of the industry continues, and even Mr. Noelke, who is better equipped than most to survive, is considering a reduction or phaseout of his sheep operation. "The sheep business can't carry one more handicap," he declares. "They can't keep piling it on us."

While typical in most respects, he is also something of a hybrid among sheep ranchers. The solidly built, blue-eyed six-footer is a 1950 graduate of the University of Texas, where he studied government and history. As a third-generation rancher, he manages about 4,000 head of sheep, and 400 cattle on a 30,000-acre family partnership. Although mineral rights and cattle sales generate some revenue, sheep historically have provided the partnership's staple income—and Mr. Noelke's biggest worries.

"For years and years, sheep carried the cows and kept us going," he remembers. That situation began to reverse itself about four years ago, and now the profit picture for sheep causes him to contemplate their phaseout and a switch of emphasis to raising cattle. "Even during the best years, we've netted $1 or $2 an acre," he says of the 30,000-acre, four-family partnership. And during the drought years of the 1950s, the spread operated at a loss. "I knew I wasn't doing this for the money," he observes.

While drought hasn't hampered ranchers much in the past five years, other problems abound. Coyotes, not yet a major problem in this part of the state, are migrating westward, inflicting increasing losses on the vulnerable flocks, ranchers say. Toxic plants, disease, drought, cold weather and brush control also keep Mr. Noelke and his three ranch hands busy.

The litany of problems is familiar to the 45-year-old rancher. His father and grandfather ranched the same land and faced the same struggle. (Their weathered saddles are still mounted on wooden sawhorses in the ranch house, and Mr. Noelke's father is buried on his land, atop a grassy ridge next to two palmetto trees.)

Mr. Noelke and his wife Margie have raised a daughter and seven sons since they were married after college. Besides ranching, he writes a regular column, "Short Grass Country," for the West Texas Livestock weekly newspaper. He also is a member of the local school board and speaks to livestock associations around the country. His friends and neighbors call him "Tuffy."

But his main job is raising sheep. "I'm a sheepman," he says. "My dad grew weary of it, but he kept them because of me. My defense of sheep is that they are the best drought insurance we have. They have saved us economically since my grandfather started this ranch. They've got a strong hold on me or I'd be out of it."

Like farming or any other form of livestock ranching, raising sheep is a year-round job. Fall is the breeding season for ewes, and some lambs are shipped to markets. During the winter, Mr. Noelke and his herders move half the flock to small pastures to feed them and keep them from eating poisonous bitterweed. Some shearing also occurs in December. Ewes begin to bear lambs in February, and more shearing is done. Ironically, spring is "the dreariest time," he says, because of the strain of waiting for the first rainfall. More shipping and shearing are done in late May, and the flock is checked on horseback and treated for disease throughout the summer.

Mr. Noelke resembles a West Texas Will Rogers in his drawling, often droll approach to a crucial time for the industry. He is particularly sensitive about antagonism between ranchers and environmentalists. "We used to be like brothers," he says. "Then somewhere along the line we had a family fight, and it turned into a big battle. Now I don't think we'll ever get back together. We need an environmental program, but we also need to make a living."

He is also concerned about who will eventually take over the ranch. His daughter is a medical student at the University of Houston, and three of his sons are studying in different Texas colleges. "Why would my son Fred come back when he graduates from the University of Texas and work a ranch under all these limitations?" he asks. "How much is it going to take for us to force people to let us produce food for them?"

Economics aside, a typical day for a sheep rancher is a world away from that of the average clock puncher or straphanger. One recent morning Mr. Noelke rose about 5 a.m. and drove his pickup truck to the ranch, about 10 miles from the family's rambling stone house in Mertzon. Blue quail and mule deer stare at the truck from stands of juniper and mesquite along the road. At the ranch, he leads his sorrel mare, Linda, to a corral and loops a bridle over her. "Used to be, there were five or six of us at a time doing this every day," he says. After saddling the horse, he loads her onto a trailer and drives to meet Felix and Jose, two hands who have worked on the ranch for more than 20 years.

"Buenos dias, senores," he says. The two mounted men nod greetings, and soon all three ride down into a brush-filled draw, searching for some 360 ewes and rams. The sheep need to be "drenched"—treated for internal parasites—and somehow have escaped treatment earlier. Fortunately, most of the flock is in one nearby pasture, and as the horsemen approach, they retreat until they stack up against a fence. Then it is a matter of bunching the animals, riding behind them and driving them to a small pen with a handling chute about a mile away.

As the herders ride, they talk quietly to one another in Spanish, beat their coiled lariats against their chaps and whistle to keep the sheep moving. "Some people ask us why we don't use walkie-talkies," Mr. Noelke says. "Why should we when we know each other's movements so well? Some other operations use herders on motorbikes. When I start having to ride a motorcycle, I'll start doing something else. I have to get some fun out of this."

As the sheep near the pen, they mill in a woolly circle, and a few try to break from the rear of the flock. Mr. Noelke wheels his horse and chases one leaping ram back into the circle. Gradually, like molasses being poured from one jar to another, the sheep plod into the pen.

The morning sun hasn't yet dried the dew on the grass, and the hoofbeats are muffled. As he walks his horse, Mr. Noelke pushes his battered cowboy hat back on his forehead and gazes across his land. "I don't think it's greed or money that keeps a man here," he says softly. "It's a morning like this."

—MIKE THARP

Hurtin' for Ground

Doyle Greenwell was "hurtin' for ground." So one Saturday morning in early 1973 he drove his pickup to attend a farmland auction in the eastern Illinois town of Marshall. He wanted that farm, or at least a sizeable chunk of it, to go with his own 300 acres nearby. He told his wife he would bid as high as $600 an acre to get it, even though a nearby farm sold for just $516 an acre not long before.

But four other farmers were also after this land and Mr. Greenwell's $600 bid was quickly surpassed. He swallowed and offered $615. Somebody else bid $625—and got the land. Mr. Greenwell climbed back into his pickup and headed for home, still hurtin' for ground.

Similar little dramas are frequently played out across the Farm Belt as farmers, flush with record net incomes in 1972 and again in 1973 and an age-old itch to expand, pushed farmland prices upward at the fastest rate since the Korean war years of 1950-51.

The land Mr. Greenwell was after sold for at least 10% more than it would have a year earlier. The overall gain in Illinois was 12% in 1972. A 12% increase in Iowa in 1972 resulted in the largest yearly dollar increase—$45 an acre—in that state since the booming 1920s. Farmland across the country increased an average 10% in 1972, and has risen further since then. That's especially notable because prices were actually declining a bit during the 1969-70 recession and rose only 5% in 1971.

The biggest increases are being paid for land to grow soybeans, corn, wheat and cotton—all crops that shot up in price in 1972 and 1973.

Forest Goetsch, president of Doane's Agricultural Service Inc., a St. Louis farm-management consulting firm, says he heard of the same piece of land selling twice on the same day—with the second sale for $450 an acre more than the first. L. C. Jones, a Decatur, Ill., broker, tells of a farmer who angrily refused to pay more than $1,000 an acre for some land, charging the seller had reneged on a pledge to sell for that price. When the seller put the farm up for sale with sealed bids, the would-be buyer suddenly handed in a bid of $1,170. He lost again; the land sold for $1,175 an acre.

Not all sales are consummated in heated bidding. Recently, Jack Koch casually mentioned in the Colo, Iowa, coffee shop that if he could get $1,000 an acre for 80 acres, he'd sell. "Why, you can get that easy," said another customer who happened to overhear, and he suggested a friend who was looking for some land. The next day, Mr. Koch happily sold and retired from farming. "But he didn't get what we could have got for him," says a local farm broker. "We've got all kinds of buyers."

Rising farmland prices even help farmers who don't sell. Land typically makes up two-thirds of a farmer's total business assets, so gains in value often mean significant improvements in the farmer's net worth and in his ability to swing operating loans and other business deals. For example, soaring land prices boosted Mr. Greenwell's net worth by at least $16,000 in 1972 alone.

Presumably it became easier for him to borrow money for, say, a new combine, if he needed one. "But I've already got $50,000 worth of equipment," he said, which is enough to farm 600 or 700 acres instead of 300. A farm twice as big as he has would be a lot closer to what experts say he needs to get the best return on his investment.

"Most farms are still too small" to maximize profits, a Chicago banker says. He thinks farmland prices will continue to be pushed up by expansion-minded farmers, who generally outbid the man trying to buy the land as a separate farm. The Department of Agriculture says that almost 60% of all farm properties are purchased as add-on land; less than a third are bought to be used as separate farms. Twenty years ago, less than 30% of all agricultural land sales were for farm enlargements.

But expansion is getting too expensive in some areas, so farmers are selling here and buying there in what brokers call the "ripple effect." A farmer simply sells his land for an attractive price and buys less expensive land elsewhere. The ripple effect helps boost prices because the big profit on the farmer's sale presumably makes him less reluctant to pay handsomely for the new land.

"Quite a few" Illinois farmers are dickering for less-attractive land in southern and eastern Iowa, says John Lowenburg, regional manager of Doane's Agricultural Service Inc. Some of the Iowa farmers are, in turn, invading the South with farmland money, according to Albert Campbell, vice president of Thompson & Associates, a Memphis-based farm real estate agency.

Perhaps surprisingly, demand for farmland from such nonfarmers as subdivision developers and persons wanting country homes isn't a major factor. Speculators near some cities are pushing prices to 10 times what the land would bring if bought for farming, but sales of land for agricultural use accounts for about 85% of the dollar value and 90% of the acreage involved in all farmland transfers, the Agriculture Department says.

Nor are corporate "agribusinesses" playing much of a role; publicly owned corporations participate in about

1% of the total farmland transfers and in fact are selling about as many acres as they are buying, a Department of Agriculture report said in 1973. There are signs of growing interest in farmland by doctors, lawyers and other absentee landlords, but their role is still relatively small.

Some analysts think the farmland boom could fizzle fast. But so far they aren't thinking that way in Farmer City, Ill., where, as the name indicates, the people take their farming seriously. They had a land auction in Attorney Ortheldo A. Peithman's office, and 30 people showed up. One was a representative of the Illinois Power Co., which was buying land in the area for a nuclear power plant. He finished a respectable third in the bidding, losing out to two farmers who said they were hurtin' for ground.

—JOSEPH M. WINSKI

Moo U.

"A good friend of mine who goes to the University of Iowa, the big liberal-arts school in our state, still likes to ask me how things are going at Moo U.," says Darryl Sywassink, who in 1974 was a dairy-science major at Iowa State University in Ames. "I just smile and tell him things are fine and that he needs a new joke."

Not long ago such ribbing might have embarrassed young Mr. Sywassink and his fellow agricultural students, who seemingly alone among college students often were challenged to justify their presence on campus. "There were always people who wondered what farmers' sons were doing in college studying farming," says Louis B. Thompson, an associate dean of the college of agriculture at Iowa State.

"But now there's increasing alarm about world hunger, and we've discovered that we can't be so complacent about food supplies in the U.S., either," Mr. Thompson continues. "So I suppose people are happy that schools of agriculture are around to teach farming. The ironic part is that for years we've been doing a lot more than that."

Tradition no doubt will dictate that students from the University of Michigan will continue to moo across the football field at their counterparts from Michigan State University and that Ohio State University, home of the country's largest school of agriculture (with a single-campus enrollment of 3,090 last fall), will continue to be dubbed "The Big Farm" by students. But the notion that schools of agriculture are cow colleges teaching farm boys what they could learn better back on the farm is becoming as passe as the horse-drawn plowshare.

Agricultural colleges point to their enrollment trends as proof that, far from being lethargic, they have been among the most progressive of all educational institutions in adapting to the changed directions of society.

From 1963 through the fall of 1973—a period in which the number of U.S. farms dropped 20%—agricultural enrollment in the member institutions of the National Association of State Universities and Land-Grant Colleges more than doubled. The figure rose to 72,644 undergraduates from 34,952. Total enrollment in these same colleges and universities rose 85% in that period.

Prior to the spurt that began around 1963 (enrollment had dropped sharply and then leveled off during the 1950s), the vast majority of students in agricultural colleges had farm backgrounds. But that has changed also. Of 1973's incoming class at Purdue University, for example, 60% had urban backgrounds and 40% came from the farm; that's the exact opposite of the ratio just five years earlier. In addition, about 20% of the undergraduates in schools of agriculture were women in 1974; previously, the number of female students wasn't deemed significant enough to count.

"During the 1960s, young people developed a tremendous concern for the environment and all our natural resources," explains Robert W. Hougas, associate dean of the University of Wisconsin college of agriculture. "When they want to translate that concern into a course of study, they find out the expertise they seek is right here."

That's because agricultural colleges have acted quickly to offer that expertise. They have beefed up or added courses such as community development, recreation-resource management and forestry that seem to match growing student interest.

"Half of our curriculum didn't even exist 10 to 15 years ago," Mr. Thompson of Iowa State says. In 1974, Purdue offered its agriculture students 28 study programs, 14 of which were listed as "new or unique," such as wood utilization and turf management. And the most popular option at the University of Wisconsin college of agriculture that year was landscape architecture.

The expansion by the agricultural colleges in some instances has sparked squabbles with other departments, such as chemistry or zoology which feel that their territory has been invaded. "Our new programs are for the most part a natural outgrowth of our acknowledged traditional expertise," argues Orville Bentley, dean of the agriculture school at the University of Illinois. "We're talking about future land-use schemes in this country, and I can't conceive of agricultural schools not playing a key role in that."

All this isn't to say that agricultural colleges no longer teach agricultural economics, agronomy or other traditional courses. Indeed, officials say that much of the 22.5% growth of agricultural colleges from 1972 to 1974 reflected the abrupt turnaround in the U.S. from food surpluses to shortages and the consequent touting of agribusiness as the growth industry of the 1970s.

For it is agribusiness—that vast network of corporations and cooperatives that supply the farmer's production needs or process his output—that continues to be the major employer of agricultural-college graduates. Fertilizer companies, grain merchandisers, food processors, farm-credit organizations and the like employ about 30% of each year's graduates, one survey shows.

"We've been besieged by recruiters these last two

years," James Mohn, placement director at Iowa State's college of agriculture, said as the 1973-74 school term drew to a close. "Our people are getting some very nice job offers." One 1974 Iowa State graduate, for example, had his pick of eight specific jobs, offering starting salaries of up to $12,500 a year. The average starting salary dangled in front of 1974 graduates of farm-belt agricultural schools was $9,200, up 12% from the spring of 1970.

Increasingly, however, the companies that court agriculture students are going away empty-handed, as another opportunity beckons: farming. "Many students here aren't even going to any job interviews," says Richard Daluge, placement director at the University of Wisconsin. "They say they can't get any better deal than they can get back on the farm." About 23% of 1973 graduates in 14 farm-belt agricultural colleges have gone into farming, up from 9% in 1969.

"Some of these students will tell you they like the life-style of farming or are concerned about feeding a hungry world, but $5 wheat and $8 soybeans haven't hurt a thing," says D. C. Pfendler, an associate dean of agriculture at Purdue.

The seed from which today's agriculture colleges and, indeed, much of the nation's public higher education system have grown was planted by one Jonathan Baldwin Turner in the 1850s. Mr. Turner, a bearded Easterner who rebelled against his classical education at Yale University, demanded in fiery speeches that universities be created to "teach the sons and daughters of farmers and mechanics."

Much of what Mr. Turner sought was provided for when President Lincoln signed the Morrill Act in 1862. That act, written by Sen. Justin Smith Morrill of Vermont, provided an endowment of public land to each state that would establish agricultural and mechanical colleges.

Today there are 71 land-grant colleges in the U.S., including one in each state and in the District of Columbia, Guam, Puerto Rico and the Virgin Islands. In addition, 17 Southern and border states have second land-grant colleges, whose enrollment is largely black; they were created in an amendment to an 1890 act to continue funding the fledgling land-grant college system.

The Morrill Act was followed by the Hatch Act in 1887, which granted federal money to the schools for research. Then came the Smith-Lever Act in 1914, which called for the schools to pass the benefits of that research on to farmers and others through what has come to be called "extension service."

"It's the oldest form of revenue sharing there is," says Purdue's Mr. Pfendler, who describes the result of the three acts as "the greatest revolution in the history of education—knowledge wasn't to be pursued for its own sake but for a purpose."

Some critics, however, charge that the agricultural colleges have strayed far from their original purpose. One example was the book "Hard Tomatoes, Hard Times" by Jim Hightower, director of the Agribusiness Accountability Project, which is based in Washington, D.C. He contends that the land-grant college system, specifically through its research and extension activities, has served agribusiness at the expense of farmers and other rural residents, driving them into the cities where they in turn add to urban problems. "Things are falling apart in rural America," Mr. Hightower charged, "and the land-grant colleges have done little about it."

But one of the more prominent products of the land-grant colleges, Agriculture Secretary Earl L. Butz, a Purdue graduate, is quick to defend the system. Mr.

Hightower's thesis, Mr. Butz once told a Senate subcommittee holding hearings on farmworkers in rural America, "is that there is something good in having a large number of people on relatively small farm units in the country. I reject that thesis." The chief purpose of agricultural colleges, Mr. Butz says, is "the development of a more efficient agriculture."

This view is held almost universally by agricultural-college officialdom, although some schools of agriculture indeed have developed programs oriented more toward people than toward production. One is Michigan State University's center for rural manpower and public affairs. "But the pendulum has clearly swung back to production," says Gary W. King, director of agricultural programs at the W. K. Kellogg Foundation, which finances agriculture-related projects at land-grant colleges, including the Michigan State project. "There's no question that the events of the last few years all have worked to reemphasize the importance of production agriculture to society and to the economy."

Not surprisingly, agriculture students are enjoying the limelight. "We used to be the dumb guys in clodhoppers and white socks," says Jay Townsend, a Purdue agricultural-economics major, who is the student representative on the university's board of trustees. "Now I sense a change—people aren't laughing at us anymore."

—JOSEPH M. WINSKI

Auction in a Corn Field

It is a March day in 1972. In a bleak corn field north of Bloomington, Ill., farmers in overalls and muddy boots crowd around a red pickup truck. Near the truck sits a piece of farm equipment. From the back of the truck comes an auctioneer's flurry of words: "I'm-a-bid 450, do I hear 475? Now I got 75, you're gonna buy it at 500 . . . all done, gonna quit, last chance . . . sold for $500."

The crowd follows the truck as it rolls slowly over the corn field's deep ruts to the next piece of equipment. "Now, here, boys, is a real nice piece of machinery. . . ."

Behind the auctioneer sits Paul Bates, 42, who has farmed this land all his adult life. He watches intently as all his tractors, trucks, and tools go one by one to the highest bidder. Finally, with the last pound of the auctioneer's fist, he is a farmer no more.

"I'd always figured somebody would take care of this after I was dead," he says quietly. "I never thought I'd ever leave the land and have to see them sell off the things I've used for, well, it seems like all my life." This is the first spring in 100 years that a Bates isn't planting this land.

Auctions like this across the country play a key role in the move from farm to city, a move that still changes the lives of many Americans every year. In 1972, the United States was losing 123 farms a day, down from the pace of a few years back but still high. Only 2.83 million farms operated that year, down nearly a million from 3.7 million a decade earlier. Some farmers retire. Some, like Paul Bates, find better jobs in town. Others just go broke. Whatever the reason, the auction is usually the means by which they leave the farm.

Other farmers are willing buyers of the machinery, livestock, furniture and sometimes even acreage itself that are sold at such auctions. Prices are often lower than on other markets. Besides, most states don't charge sales tax on items bought at such auctions. An estimated $300 million worth of machinery and equipment alone changes hands at farm auctions each year.

But for all its bargains, the farm auction, like the small family farm itself, is a fading piece of Americana. The number of auctions peaked during the Depression when bankers often held them as part of foreclosure proceedings. Sympathetic neighbors often agreed to bid only a tiny amount on each item, then give all the goods back to the farmer being squeezed.

The number of auctions in recent years is dwindling mainly because there are simply fewer and fewer small family farms left to be sold. In the area around Bloomington, the number of auctions in 1972 had declined about 30% from the level three years earlier, says Stanley Lantz, farm editor of the Bloomington Daily Pantagraph. "Maybe within this decade, we'll see farm auctions become a thing of the past," he says.

Auctions are important social and business events in a farm community—and emotional events for people like Paul Bates. His situation is common on American farms. "I'm a little bitter about quitting farming," he concedes the night before the auction. "The circumstances aren't entirely of my own making, but my better judgment tells me I'd best get out now. I know it's no use to get emotional about something like this—but, God, I wish I didn't have to be there tomorrow."

To supplement his farm income, Paul in 1971 took a job as a commodity futures broker with the Illinois Agricultural Association, a trade group that merchandises grain, among other activities. Paul already had some experience in grain trading, and he had worked part-time in winters on farmers' books. But even with the new job, he continued farming. Though he quit the hog-fattening business, he found he could manage his own 270 acres plus about 400 rented acres by working nights and weekends raising the corn.

Paul planned to do the same in 1972, but on Jan. 7 of that year a fire destroyed his barn, biggest tractor, biggest truck and considerable machinery.

"We were insured, but we got only $6,500 on the tractor, and it would cost about $13,000 to replace it," Paul said. "So it came down to whether we wanted to invest about $30,000 to keep going or pull out. Considering that we made only about 3% on our investment last year, and that my salary is bigger than the cash I made by farming, pulling out looks like the sensible answer."

Paul, who is divorced, will still live in the wood frame farmhouse with his 17-year-old daughter. His 65-year-old father and farm partner, Tom Bates, will also continue to live in a separate house on the farm. Since heart trouble laid him up for five months in 1969, Thomas Bates has steadily reduced his farm work. He clearly was anxious to quit farming altogether and relax in his sedate job as a watchman at a nearby orphanage. Unlike his son, he was tired of the farm's responsibility. By the time of the auction, the Bateses had rented their land to a neighbor in return for half the crop.

The afternoon before the auction, Paul sits at his kitchen table with the auctioneers who come to size up the goods they are supposed to sell. The auctioneers, Kenneth Coulter and Dean Yoder, are professionals at the work, though they also keep a hand in real estate and farming. In their appearance, they are a combination of farmer and mod city slicker. Mr. Coulter wears a bright green sports coat, pink shirt and pink tie—along with cowboy boots and Western hat.

The two auctioneers breeze into the kitchen, laugh loudly at their own jokes and slap Paul on the back. "I guess that's their bedside manner when they see a guy like me feeling a little down," Paul says later, with appreciation.

The auctioneers discuss prices and warn Paul he is in for some disappointments. "The prices you'll get for some things will be so low your feelings will be hurt, but other items will bring a lot more than you figured," Ken advises.

Auction day itself is sunny, but chilly. Three hours before the 11 a.m. starting time, farmers begin arriving to inspect the goods. They poke and pry. They drive the tractors. They ask questions:

"How long you had the planter, Paul?" "Do the gas pumps work?" "How could I hook up that John Deere mower to my International (Harvester) tractor?" Paul and his father patiently answer each one.

The two auctioneers arrive in their red truck and begin to register bidders. Each of the 216 registrants receives a number to flash at the clerk if he buys something. "It speeds things up quite a bit to use numbers instead of names when we sell," Mr. Coulter explains. "We know a lot of these men personally, but if we see a stranger buying up a lot of things, one of us will slip into the house and make a telephone credit check with the guy's local bank while the auction is still going on," he adds.

In the dealing itself, a country auction demands its own brand of shrewd sophistication. "The real skill of an auctioneer is to figure out just how far the bidders will go and then make sure they get to that point," Mr. Coulter says. The purpose of the auctioneer's fast patter is to build excitement and a semblance of momentum, even though the bidders themselves may actually be pausing to think things over. A dishonorable—but widely used—auctioneer's trick is to throw in a few fictitious bids to prod the actual bidders to higher and higher levels.

Some items are even sold fictitiously. The seller usually gets his own bidding number, which he keeps secret. Thus he can "buy" slow movers that are dampening the crowd's enthusiasm or items that aren't bringing high enough bids. For instance, Paul keeps number 44, sits in the truck with the auctioneers and discreetly taps one of them on the shoulder when an old wooden picnic table attracts a top bid of only 50 cents, Paul "buys" it for 75 cents.

The auctioneers' skill apparently helps boost prices on several items. A hay mower that Paul bought second-hand for $75 eight years earlier brings $90 in spirited bidding. A 1960 Chevrolet truck that Paul figures is worth about $1,000 sells for $1,600. On the other hand, an old planter sells for $700 instead of the $1,000 that Paul expected. Another piece of equipment that might command over $700 in a used-machinery store goes for only $370.

One reason for these low prices, of course, is that buyers have tricks, too. For instance, one savvy farmer who frequents many auctions deliberately tries to slow down the pace. If the auctioneer wants to open bidding on a tool at, say, $5, this farmer may bid 75 cents. He raises his bid only 25 cents a time and eventually wears out his competitors.

"This guy knows a lot of people will bid just so many times on something and then quit, even if the price is way too low," says Mr. Coulter with grudging respect. "I've seen him pick up a lot of power tools and stuff for chickenfeed."

An elderly farmer with a cane uses another hoary

trick. Long before Paul's auction starts, he hangs around a small tractor. When other farmers stop to look at it, he sidles up and shakes his head. "Needs new brakes," he says knowingly to one man. "Not in such good shape," he murmurs to another. But when the bidding gets around to this tractor, the elderly man bids vigorously against all comers. He smiles broadly when he is declared the buyer.

After four hours of selling, the auction reaches its finale; bidding for the big combine. On such big-ticket items, the auctioneers openly beg and cajole. "Come on, fellas, you know darn well you'd pay twice as much to a dealer for a machine in this good a shape," Mr. Yoder pleads. "Paul will never speak to me if you don't give me at least three-quarters of what it's worth." Paul had hoped to get $7,000 or $8,000 for the five-year-old piece of equipment. The auctioneers considered $6,500 more likely and $7,000 the absolute top. The combine goes for $6,900.

The farmers quickly take charge of their purchases and settle their bills. Payment is by cash or check. Bills are settled in the kitchen where the auctioneers' wives have set up shop with two adding machines and a cash box.

Within half an hour, nearly everybody is gone, and the auctioneers' wives total the proceeds. The auction brings in nearly $26,000, which Mr. Coulter says is a little more than average. It's about what Paul had hoped to get. The auctioneers collect their 2% fee; the clerks and cashiers share 1%.

With a cheery beep-beep on the horn, the auctioneers speed away. It is over. Paul Bates is an ex-farmer, yet another country gentleman who lives in a farmhouse but works in a town. He knows things will never be the same.

"Farming is a great way of life—until it stops being

a way of life," Paul muses. "I've worked so hard at it that I never learned to swim or dance or play golf or any of the things that other people do to enjoy themselves. I always thought, well, I have my farm. But now I have nothing, except my job. It's a very lonely feeling."

—JOHN A. PRESTBO

Part Two

HOW FARMERS FARM

There is a lot more to farming than planting and harvesting, and more to livestock raising than feeding the animals and shipping them off to slaughter. Modern farmers use a variety of techniques and methods, which are applied in a variety of places—not all of them in the countryside. They have harnessed computers and airplanes in an effort to boost production. They have banded together in marketing groups and have tried selling direct to consumers in an attempt to influence the prices they receive for their products. Farmers are, in fact, constantly trying new wrinkles, which makes U.S. agriculture ever-changing and progressive.

Farmers in the City

Los Angeles is definitely no place for an average farmer, especially if he worries too much about little things like corn borers, hailstorms and hog cholera.

In Los Angeles, a man of the soil has problems that most of his Corn Belt colleagues never dream of. Like smog. Sometimes, when it's thick enough, your whole radish crop can get zapped.

Then there are the highwaymen. If your oranges are ready to pick, a string of passing motorists may well have noticed, too; by the time you get to the trees, the motorized bandits are already en route to Anaheim, Azusa or Cucamonga, auto trunks crammed with a month's supply of breakfast juice.

Even juvenile delinquency is a problem. One farmer carelessly left two tractors in his fields after the day's work. At midnight he found a couple of local hot rodders playing a game of chicken with the machines; they were driving full speed toward head-on collision, the more fainthearted veering off at the last instant.

Such is the farmer's lot amid the freeways, factories and more than seven million residents of Los Angeles County, the nation's most populous county. Agriculture is a trying business there, and continuing urbanization and rising land values, which induce farmers to sell out, are making it a dwindling one. Still, there are enough farmers remaining to rank the county quite high in the U.S. in the value of its agricultural production.

Farms can be found just about everywhere in the county, whose 4,069 square miles make it more than three times the size of Rhode Island. Some are conventional spreads in still-rural parts of the county. But many others are next to oil refineries, under power lines, snuggled in the forks of freeway intersections and surrounded by subdivisions. In places like these, a farmer has to be adaptable to survive.

Roy Pursche, for example, tries to outwit the smog that often shrouds a lima-bean patch he tills next to an oil refinery. He plants earlier than usual, so he can harvest before the heavy smog buildup common in late fall, and he has arranged his rows in line with prevailing winds so they can most effectively blow the pollution away.

But smog is a tricky foe. "It can make a plant look like a blowtorch was passed over the leaves," says one agriculturalist working 127 acres leased under power lines, hard by the Long Beach Freeway. A county farm official reports that a two-day siege of heavy smog once ruined $200,000 worth of crops in the county.

The local farmers must also be adept at public relations. Considering that their cows and chickens produce 25 million cubic feet of manure a year, it's not hard to understand why; the fragrance so familiar to farm boys in the Midwest is a blight in the nostrils of the split-level set, especially when backyard barbecues are under way.

Some canny farmers enclose their property with fences that hide it from view. "When the residents can't see it, they seem not to be able to smell it," says Kenneth M. Smoyer, county farm adviser. Other farmers take more positive action. Mr. Smoyer tells of one poultryman who builds goodwill by selling eggs in his neighborhood wholesale and offering the same deal on turkeys and chickens at Christmas. "It's interesting," says Mr. Smoyer, "that there are no odors at that man's farm and no trouble with neighbors."

While winning a few battles, however, dairymen

and poultrymen seem to be losing the war. Health authorities, prodded by complaints from local citizens, watch them like hawks, forcing prompt disposal of manure. Ordinarily, that wouldn't be much of a problem, but in recent years the market for raw manure has collapsed. Norman Lautrup, a La Mirada dairyman, has to pay someone to take his away, whereas he used to sell it.

Mr. Lautrup is also a little miffed at area youngsters. He says one group burned up one of his haystacks by shooting lighted matches into it, and he reports that others like to chase his high-strung Holsteins around the pasture and throw rocks at them, hoping to get them to jump the fences. Nervous cows don't give as much milk, he complains.

Some suburban communities have decided to give Bossy the heave-ho. Elsewhere, dairymen who once operated in relatively pastoral settings now find themselves ringed in by homes—and besieged by complaints.

Most farmers here seem philosophical about the problems, however. Thieves are taken in stride, so long as their needs are modest. "I don't mind a hungry man getting something to eat from my field," says George E. Giardina, a vegetable grower, "but some of these people try to fill the trunks of their cars and some even try to sell what they've stolen."

Citrus growers sometimes keep the outside rows of their trees closely picked in order to discourage thieves, but now they have a new problem—highwaymen who dig up whole trees and cart them away. For sheer gall, however, it would be hard to top the fellow who sneaked a combine into a barley field of Roy Pursche, the man who is trying to grow limas next to the refinery, and began harvesting his crop. He covered 10 acres before Mr. Pursche nabbed him.

For all the problems, though, a city farmer can

make a lot of money. Consider Louis De Martini Jr., a big man in the local soul food business. He grosses $250,000 a year selling collard and mustard greens to chain stores in the predominantly Negro south central area of Los Angeles.

Mr. De Martini's land is only six miles from his customers, meaning grocers can order as late as 8:30 a.m. and get delivery of fresh-picked greens the same day— service that farmers outside the city can't match, Mr. De Martini says. These advantages outweigh some of the hazards of the trade, such as the time he found himself and a truckful of greens caught in the middle of the 1965 Watts riot.

—BYRON E. CALAME

Hamburger on the Hoof

When Joseph Ference was growing up on a Connecticut farm, he occasionally raised steers for neighbors. The neighbors got their beef far under retail prices, and young Joe made some spending money.

In 1974, Joe Ference was a husky 32-year-old, but his boyhood hobby was once again very much alive. He was raising 86 steers on his rolling, 247-acre dairy farm outside the tiny northern Vermont village of East Hardwick for his customers—Boston and New York suburbanites.

Mr. Ference's venture is part of an increasing tendency by farmers to trim consumer costs and fatten their own profits by selling meat directly to consumers. For a flat fee of $600 a steer, Mr. Ference not only raised the animals and arranged slaughtering but also guaranteed the purchaser of each steer 600 pounds of meat —steaks, roasts and hamburger wrapped and ready for the home freezer. "I didn't see how I could buy meat anywhere else for a buck a pound," said Glen Insley, an institutional bond salesman for Merrill Lynch, Pierce, Fenner & Smith in Boston.

Mr. Ference was realizing a profit of about $50 more than he would get on the wholesale meat market. He also gets paid in advance when he starts raising the calf. Thus, he obtains additional working capital and is free of the vagaries of the beef market. Demand has been so great with just a little advertising that he plans to phase out his 44-cow dairy herd and perhaps by 1976 to be raising 500 steers annually for direct sale to consumers. He figures that his profits would increase a good 50%, to about $15,000 from $10,000 in 1974.

There isn't any way of knowing just how much meat is being sold directly from farmers to consumers. "It's been a new development in the meat industry," says an official of Beacon Milling Co., a Cayuga, N.Y., feed-grain firm. The trend is probably a natural outgrowth of the increasing tendency by city types to head for the country to buy fruits and vegetables and in some cases even pick their own apples and strawberries in the fields.

In fact, no food seems out of bounds if the price is right. In Maine, for instance, some dairy farmers find themselves deluged with people wanting to buy raw milk—unpasteurized and unhomogenized—at 90 cents a gallon. "With the food shortages and the high costs, we're going to see more of this," says Francis Nelson, product manager and nutritionist of Domain Industries Inc., a New Richmond, Wis., feed-grain firm.

Most direct-to-consumer meat sales up till now have been made by word of mouth. Mr. Ference advertised his venture in two Fairfield County, Conn., papers, but he says that most of his customers heard of him from other customers. Wilbur Pearce, a Perryman, Md., farmer, sold about 40 of his 160 head of cattle to friends and neighbors in 1974 up from 20 in 1971, all without any sales push. He sold the rest of his cattle on the open wholesale market. George Moore, a Geneva, N.Y., farmer who sold one steer to some friends in the late 1940s as a favor was in 1974 selling about 300 steers to individual consumers. Although he has never advertised, business lately has been increasing about 20% annually, he says.

Considering the fanatical loyalty shown by some customers, advertising may be superfluous. Paul Costello, a marketing manager for Chesapeake & Potomac Telephone Co. in Washington, drives 350 miles once each year to a slaughterhouse near Mr. Moore's farm in upstate New York to pick up the beef of a steer that he then splits with a neighbor. "The quality is so good we'd buy it even if the prices here in Washington matched Mr. Moore's prices," he says.

Some would-be and part-time farmers are also seeing dollar signs in direct-to-consumer sales. In the summer of 1973, Jim Nudd, an electrician and part-time farmer with 20 acres of land in Walden, Vt., sold 12 pigs to area residents at about $170 each, compared with about $100 that he would have received on the wholesale market. The sales were so easy, he says, that within a few years he expects to be out of electrical work and to spend all his time raising 600 pigs a year for direct sale to consumers.

Not that raising meat for consumers is a surefire way of making money. Ask Reinhold Pretsch, a Plain, Wis., farmer, who in 1973 bought 3,500 roasting chicks to raise and sell to area residents. Mr. Pretsch found that while his chickens sold well, feed-grain prices were so high that he couldn't make any money on the chickens.

Mr. Ference, the Vermont farmer, has taken great pains to cover himself against that and other problems. For instance, when a customer orders a steer, Mr. Ference immediately assigns a calf to the individual and orders enough feed grain to raise the calf for the one year it takes it to become a 1,000-pound-plus steer. He knows his costs in advance, and "customers get a break because they get next year's meat at this year's prices," Mr. Ference says.

He also guards himself by personally feeding, cleaning and inspecting the animals—in the course of 14-hour and 16-hour days—to prevent them from getting sick. To avoid being caught short by a dying animal, he always keeps a few more calves on hand than he has sold. He either buys calves from neighboring farms for about $60 each or breeds his own, using two bulls he owns.

Most of his urban and suburban customers know nothing about cattle, Mr. Ference says. "Some people think you just go down the animal from one end to the other and cut steaks," he says. They are unaware that 450 pounds of a 1,050-pound steer is waste from the head, skin and blood, Mr. Ference says. And, he adds, they don't realize that the 600 pounds of meat includes not only the prime ribs but also the stewing beef, tongue and liver.

A few customers make special requests. One suburban New York woman, for instance, asked Mr. Ference if she could watch her steer be slaughtered and have the hide to make into a skirt. He told her she would have to arrange with the slaughterhouse to see the butchering and buy the hide. The slaughterhouse keeps all remains as part of its payment (in addition to the slaughtering fee of about $50 an animal charged Mr. Ference). Another suburbanite wanted the steer's head to feed to his dog.

Mr. Ference invites customers to visit his farm, to view their animals, but he has few takers. His wife, Patti, thinks she knows why. She rarely ventures out to the nearby barn to see the steers because, she says, "I can't stand the idea of eating an animal I've named or grown to like."

—DAVID GUMPERT

POSTSCRIPT

When this article first was printed in The Wall Street Journal, two readers in Millbrook, N.Y., and Erie, Pa., both familiar with the cattle business, wrote letters saying it couldn't be done. Well, when Mr. Ference slaughtered his first four steers and delivered the meat to customers in Boston, it looked like those two readers were right.

Live weights of the four steers were 1,100 to 1,200 pounds each, Mr. Ference said. After being skinned, de-headed, drained and cooled, the carcasses weighed 598 to 605 pounds each. The butcher who cut up the steers said there would have been a further weight loss of about 10%, or 60 pounds, from normal cutting and trimming, except that the customers wanted lean meat, boneless roasts and no tongues. So the losses from trimming and cutting ran 16% to 18%, or roughly 100 pounds a steer, the butcher said.

Although Mr. Ference's customers got only about 500 pounds of beef for their $600 they don't seem too unhappy. Mr. Ference offered to make up the difference from steers he has been raising for himself if any customer thought he had been shortchanged. But he hasn't had any takers.

"Based on the price of meat today, I think we got a good price," one Boston customer said. Another, John Foster, said, "It's a good value." He said he was "disappointed," but "I don't think there was any deception." Nils Peterson, a third Bostonian, was sufficiently satisfied to order another steer from Mr. Ference.

Oddball Crops

Every so often a stranger will drop by the Gumz farm in northwest Indiana to pick up a bottle of home-grown "essence," which many of the local folks swear is a good remedy for sore muscles and upset stomachs.

Owner Steve Gumz says he doesn't know if the stuff works, nor does he really care. What Mr. Gumz does care about is that the mint oil he distills from his 550 acres of peppermint and 150 acres of spearmint— one of the largest mint farms in the country—brings the premium prices he feels he deserves after a lot of hard work.

Thanks to burgeoning domestic and overseas demand, Mr. Gumz is doing nicely. Since 1972, demand from makers of toothpaste, mouthwash and chewing gum (mint's primary customers) has pushed mint prices to $15 a pound from $5 a pound. As a result, more and more farmers have turned their soil over to mint. The 1974 harvest was expected to be worth about $60 million, up from $35 million in 1973.

Viewed against a $14 billion corn crop and a $9 billion soybean crop, mint doesn't seem to be a very important part of American agriculture. But mint is only one of 200 or so specialty crops raised in the U.S.; taken together, the so-called "oddballs"—including such crops as cocktail onions, horseradish, dill and ginseng— amount to a sizable and growing industry. Moreover, it's one of the few segments of farming that offers growers a unique advantage over their colleagues who grow more conventional crops—markets in which individual producers can swing enough weight to inflence prices.

Some of these crops, including mint, have been

grown for special markets for a long time. Interest in others has been spurred by recent health and diet trends, increased gourmet cooking in the home and a general disenchantment with conventional foods. "With grocery shelves stacked with processed, packaged and canned commodities, the one place the housewife is romanced today is at the specialty produce counter," says Freida Caplan, owner of Produce Specialties Inc. of Los Angeles.

Growing oddball crops isn't easy, though. Many of them require special soil and are very demanding upon it. Most need special fertilizers and equipment. And some are highly susceptible to diseases for which there aren't any cures.

Once a farmer starts growing an oddball crop, however, "you're more or less captured," says Cornelius Reitveld, a third-generation cocktail-onion grower near La Crosse, Ind. That's because these farmers are virtually locked to their land, having located a suitable soil, shelled out a lot of money and learned all the horticultural tricks required to grow oddballs. By the same token, it's difficult for a farmer to break into raising specialty crops.

Cocktail, or "pearl" onions, for example, insist upon a rich, moist soil, a regular rain supply and just the right amount of sunlight. Pests are a persistent problem, and only expensive pesticides can ward them off. The crop has to be rotated regularly with a conventional crop to preserve the soil. (Mint, on the other hand, so ravages its soil that the plant can be grown in the same spot for only two or three years and never again; this causes the truly dedicated mint grower to be something of a nomad.) Cocktail onions must be sold soon after harvest, or they will lose their color and sweet flavor. Tilling equipment is costly because it must be specially designed for a plant that's only six inches tall.

Similarly, raising horseradish, the root of which is processed into a pungent condiment, is a painstaking task. It involves pruning or "suckering" the plant early in the season, "lifting" it—removing the soil from around the root—during the summer and, finally, harvesting it, which requires special deep-digging machines to lift the roots, as well as much manual labor.

Even then there's a good chance the crop will be attacked by its bitter enemy, the horseradish flea beetle, an insect for which there isn't a known poison and thus is "extremely difficult to control," according to the U.S. Agriculture Department. Having survived all this, a horseradish farmer's worries aren't over; stored horseradish that is exposed to light will turn from its normal whitish color to an unappetizing (and unprofitable) shade of green.

The farmers who grow horseradish, usually under contract to dealers, were rewarded in 1974 with rising prices. The spurt was partly due to a short supply in 1973, when some growers cut back production because of then-unsatisfactory prices. The 1974 crop was estimated to be worth a record $2 million, a sizable sum when divided among only 55 or so growers throughout the country.

"This is a business that requires a degree of sophistication, so not everyone who tries it succeeds; that's why there are so few of us," says Mr. Reitveld, the cocktail-onion farmer. "It's also why we have a market that we can pretty much control." As such, it's also a market that can be quite lucrative. "If I could, all I'd grow is cocktail onions because that's where the big money is today," Mr. Reitveld adds. He can't confine himself to cocktail onions because of the need to preserve the soil through crop rotation. Thus, he also grows corn, soybeans and carrots.

A lot of oddball crops are receiving attention because of increased interest in gourmet cooking. One such crop is the Florence fennel, also known as finocchio, which resembles celery but is grown for its sweet seeds that are used to flavor salads, breads, pastries and candies.

Gourmets say that endive, escarole and chicory also make for chic additions to salads, but nobody seems sure which is which. "If you can keep them straight you're doing better than most professional produce men I know," says Joe Carcione, who writes a daily column on produce for the San Francisco Chronicle and who is considered an expert on oddball crops.

The leek, a mild member of the onion family, is being grown increasingly in the U.S. as a flavoring for soups and sauces. In France, the leek is known as the "asparagus of the poor" because it can be boiled and served like asparagus, but is cheaper. Wales, however, has held the leek in high esteem since 640 A.D., when the Welsh scored a big victory over the Saxons—at least partly because the Welsh soldiers had leeks pinned to their helmets and thus didn't attack each other by mistake. The lowly leek became the country's national emblem, and it still is.

Dill is also becoming a gourmet's delight, says Marshall Meale, a New York consultant to the American Spice Trade Association. He says the plant's aromatic seeds and leaves are added to stews, soups and seafood sauces, as well as being used for pickling and baking.

A resurgence of "soul food" has boosted cultivation of a branch of oddballs commonly called "greens" that are used in making thick soups. Particularly in the rural South, these vegetables—chard, collard, kale and mustard—often are stewed with salt pork and eaten with corn bread to sop up the "potlikker." Many people simply steam or boil the greens and eat them like spinach.

The increased popularity of Chinese food in the U.S. has given rise to the cultivation of some Cantonese vegetables here. Among them are the Chinese sugar pea, which is eaten pod and all, and Chinese cabbage, also called "napa" or "bok choy," which resembles conventional cabbage but has longer and narrower leaves and isn't quite as dense. The Agriculture Department says that about 1,200 acres of Chinese cabbage is being grown in the U.S., mainly in California.

Many specialty crops appeal to health- and diet-conscious consumers. For example, sunchokes, formerly known as Jerusalem artichokes (the name was changed to broaden the appeal of the vegetable, produce men say), are used by a lot of people as a potato substitute. Sunchokes are said to have twice as much vitamin A, thiamin, riboflavin and calcium as potatoes but 25% less starch. What's more, the carbohydrates in sunchokes, which look something like sea mines, apparently don't convert into sugar; this makes it an attractive diet item for diabetics, some people say. About two million pounds of sunchokes will be baked, boiled, fried and mashed this year, a tenfold increase over five years ago.

A type of cucumber developed in England is said to be tender, seedless and "burpless," and thus good for delicate stomachs. As a result, it has become nearly a $10 million industry almost overnight in the U.S.

And then there's ginseng, a plant that can be found in the wild but increasingly is grown commercially and is sold in health-food stores as a nutritious substitute for tea and coffee. In many countries, including the U.S., it also is used as a mysterious, talismanic medicine to treat everything from impotence to arthritis. It's supposed to be good for the common cold too, but there may be cheaper remedies on the market. High-grade ginseng sells for about $50 a pound.

—GEORGE GETSCHOW

Fish Farmers

What can a farmer do with six acres of mesquite, sagebrush and cactus on land so dry it takes 40 acres to keep one steer alive? Why, it's a perfect place to raise 22 million catfish.

At least that's what Don Carr says—and he's regarded as one of the sharpest entrepreneurs in the border town of Eagle Pass, Texas.

In 1971, the 29-year-old cowboy headed a group of Texas investors that had staked over a million dollars on an operation that includes everything from a complete hatchery to an ultramodern processing plant. Mr. Carr acknowledges that "very few people have made much money farming catfish." But he foresees the time when the "Ictalurus Punctatus" will grace tables in every section of the country. So important will catfish become to the nation that someday there will be a museum to the pioneers of the catfish industry, "just like baseball's Hall of Fame," he maintains.

Mr. Carr may be overly enthusiastic about what Mexicans term "buzzards of the river" (he also distributes bumper stickers that say, "Give catfish for Christmas"). Nonetheless, catfish farming has spread rapidly throughout the Southern U.S. and now reaches from coast to coast. Catfish connoisseurs have a hard time describing why they consider the fish a delicacy, but more and more people are beginning to share their feelings.

Nationwide output was expected to reach 84,000 tons by 1975, up from an estimated 37,500 tons in 1971 and a mere 13,500 tons in 1968. The bewhiskered fish, which derives its name from its catlike head, is being raised in ponds, troughs, tanks and cages. Earl Hansen, an undertaker in Holstein, Iowa, has even invested in a catfish building — the second such operation in the world. "We have," he says proudly, "the biggest cathouse in Iowa."

Using cages, another farmer, James Kelley of Colorado City, Texas, originated a system for growing catfish in the effluent water of a hydroelectric plant, demonstrating, he says, that so-called thermal pollution can be a bonus. He is, he maintains, the "originator of kilowatt catfish" ("the only shocking thing is the price"). Mr. Kelley claims to be able to produce 35 pounds of catfish in a cubic foot of cage space every four months. His operation is considered so unusual that a group of Japanese fish farmers came to examine it.

Profit is the main lure for catfish farmers, some of whom say catfish is more lucrative than any of their other crops. For example, Nathan A. Cormie of Iowa, La., said rice was bringing him $40 an acre at most in 1971, but with catfish he cleared $175 to $250 an acre. It cost 26 cents to produce a pound of catfish that year (feed costs have risen sharply since then) and he usually got 40 to 50 cents a pound for live fish that weighed more than one and a half pounds. Most fish are sold live to processing plants, which freeze them for sale to restaurant chains, frozen-food outlets and traditional fish markets.

Edgar Farmer of Dumas, Ark., who has 800 acres in ponds, says he sold a third of a million dollars of catfish in 1970. "We're not even touching the market yet," he says. "We can sell an unlimited number of catfish."

"Tiny" Harris, a huge, cigar-smoking man who runs a profitable gravel business and raises Brangus cattle and hogs, estimates that his 18 acres of catfish water net him $5,000 annually, roughly 10% of his income.

And apparently some hard-nosed corporation heads

figure catfish may be even more profitable than widgets. Pennzoil United dropped its hook into the business in early 1969 and by 1971 had 338 acres of ponds in Southern Louisiana. A spokesman termed the operation "hopeful." General Tire & Rubber, which had a catfish farm at Homestead, Fla., expected to raise production to seven million pounds annually from two million. Dow Chemical Co. was experimenting with cage production of catfish at Freeport, Texas.

And catfish farmers have perfected the "raceway." A raceway is to catfish what feeder yards are to cattle. Mr. Carr's Big River Catfish Farms concentrates on a raceway system involving long, rectangular pools of flowing water from the Rio Grande. Catfish can be packed closer together in raceways than in ponds, much as cattle feeder lots have replaced the open range. In 1971, Mr. Carr sold a half-million pounds of dressed frozen catfish for $60,000, and by 1975 he expected to sell 20 million pounds.

Besides the money, catfish farming has further appeal. For devotees of the leisurely life, such farming can be less than arduous. And that makes it fine for retirees and moonlighters as well as lazy people.

Like other businessmen, of course, catfish growers run into a wide variety of problems. One is depletion of oxygen in the tanks, which overnight can wipe out a catfish farmer figuratively and his crop literally.

Another is that more than just people like catfish. For instance, alligators. One Florida farmer occasionally finds 'gators gobbling up his profits. Other farmers report similar problems with otters, mink, frogs and snakes.

Then there are people—the kind that poach catfish (and not like an egg). Poachers were the biggest problem for E. W. Abdo of Kleburg, Texas. Then he bought Lady, a large German shepherd that patrols his indoor catfish channels and takes a hard-boiled approach to poachers. Another way to deal with catfish rustling was developed by a Missouri biologist. He developed a brand for catfish, much like a brand for cattle.

Another problem is Yankee ignorance of this Dixieland delicacy. In fact, breaking the catfish barrier in some Northern sections is considered nearly impossible by Southerners. Laments John Tallant, sales manager of Goldkist Fish Co., Quitman, Ga.: "New England people just won't accept damned old catfish. It's not worth beating your brains out up there."

Still, catfish connoisseurs abound, and they're a faithful lot that businessmen consider worth courting. "Our catfish customers won't accept substitutes. If you don't give them catfish, they'll throw it right back at you," declares Everett Wayne Jones, who initiated a chain of catfish franchises. "You can't call it catfish and give them codfish."

Catfish lovers give differing comments when asked to describe how the fish taste. Says the owner of a catfish restaurant in Dallas, "It's nothin' like shrimp, nothin' like lobster. I just don't know how to tell you." Most catfish eaters stress the sweet taste and say "farm-raised catfish doesn't taste anything like muddy-river catfish." They say the taste isn't fishy or too strong and has no oil or iodine taint.

Rep. Joe E. Waggoner of Louisiana declares, "There is an old saying in the South that if there is anything better than catfish, the good Lord kept it for himself."

Despite the proliferation of recipes such as catfish caper, catfish Evalin, catfish kabobs and catfish pudding, most catfish lovers prefer catfish fried in cornmeal batter in the same pan with hush puppies. (Hush puppies, popular in the South, are small, unsweetened cornmeal cakes.) Even a fancy Washington, D.C., restaurant, the Black Sheep, which boasts escargots, steak

Diane and frog legs Provencal, serves catfish "down-home style."

And according to many Southern folks, catfish farming is a pretty palatable recipe in itself. Take some warm outdoors and lazy waters, add thousands of whiskered catfish and spice it with profit and tasty eating.

"Catfish farming is living," sighs Mr. Carr. "Everything before or after is just waiting."

—DOUGLAS MARTIN

Organic Farming

"In the soft, warm bosom of a decaying compost heap, the wheel of life is turning."

This observation in a pamphlet about organic gardening may seem a bit emotional to the average weekend gardener, who thinks a compost heap is just a smelly pile of manure, rotten leaves and moldy grass clippings. But organic gardeners and farmers take them seriously. And because of the controversy over the safety of pesticides and food additives, a lot of other people take them seriously, too.

Organic farmers grow their produce without any help from chemical fertilizers, sprays, dusts or pesticides. There was a time when the market for such foods was a narrow one and the general public probably lumped both the farmers and their customers together with food faddists and eccentric old ladies in tennis shoes.

No longer. Today, ordinary housewives are joining those food faddists at the counter of their neighborhood health-food store. Meanwhile, leaders of the organic movement say biologists, doctors, conservationists and others are rallying to their banner.

There are 1,500 to 2,000 health-food stores around the country (not counting the roadside stands that some growers operate), and there are plans to launch more. In San Francisco, small groups with names like "Friends of the Earth" are snapping up organically grown produce at places like "Far-Fetched Foods," a little store in the Haight-Ashbury district.

Organic growers raise nearly all the common fruits and vegetables, along with more exotic fare like alfalfa sprout. Growers also produce organic beef and poultry from animals that eat organically grown food and never get growth-accelerating hormones. Such food is expensive. Before Thanksgiving 1969, Lindberg Nutrition Centers in Los Angeles featured organic hen turkeys at nearly 30 cents a pound more than ordinary gobblers were selling for at a nearby supermarket. But a proud Lindberg store manager defended the price. "These big, beautiful broad-breasted birds are magnificent," he said.

Organic farmers say their food costs more because they have to sacrifice the higher yields chemical fertilizers and pest-killers would give them in order to grow tastier, more nutritious food the way nature intended. Novice farmers often find their ventures far from profitable.

Take the Rev. Herbert Larson, for example. A retired Lutheran minister who says he was "in the wrong racket all my life to make money," Mr. Larson rented nine acres of land at $20 an acre near Maple Bay, Minn., bought $3,000 worth of equipment, hauled in 28 truckloads of vintage manure (the older the better, organic growers say) and planted more than 5,000 raspberry bushes.

Over three years, Mr. Larson sunk about 2,000 hours into his project. Then he sold $430 worth of raspberries. "I ate some and gave some away, and I guess I didn't charge enough for the ones I sold," he concludes.

Insects usually are the biggest problem for organic growers, who devise ingenious means to combat the pests without resorting to chemicals. Lee Anderson, an elderly date grower from Indio, Calif., for example, had a problem with date beetles. But then he noticed the beetles spent the off-season among the fruit lying on the ground. So, after each harvest he sent his herd of 150 hogs through the palms to eat the fallen dates. He also hung buckets of watermelon rinds on the trees; when the beetles were lured into the buckets, Mr. Anderson scalded them with boiling water.

Another way to fight pests is to introduce a natural predator. Thus, ladybird beetles often are brought in to devour troublesome aphids. Plant diseases are best prevented by making sure the soil is rich in humus, the decomposed organic matter formed after compost is worked into the soil, organic experts say. (In fact, almost any discussion about organic gardening eventually gets around to humus. "You must develop a sense of humus," admonishes one pamphlet.)

Organic enthusiasts say they're willing to put up with all kinds of hardships in return for their unpolluted food. Mrs. Sid Ries, a retired housewife who says her four-and-one-half-acre garden near Missoula, Mont. has "compost piles all over the place," admits that bugs are a problem. But, she adds, "I'd rather find a worm in my apple than DDT on it."

—GEORGE GRIMSRUD

Farming Without Soil

In an age of scientific marvels, a burpless cucumber might pass unnoticed by everyone but salad lovers with nervous stomachs. (It was put to the test at a cucumber dinner for a few people who love them but are usually afraid to eat them. Not one burp ensued.)

Nobody knows exactly why this European hybrid cucumber is kinder to the stomach, but the way it was grown might have something to do with it. It and other kinds of produce are raised not in soil but in trays lining greenhouse-like structures with controlled temperature and humidity, and they are fed by chemical nutrients mixed in water and automatically metered out as needed. This method of growing plants is called hydroponics.

Hydroponic farms across the country are rapidly expanding their production of tomatoes, cucumbers and a scattering of other products. Most of these outfits have begun operation within the past few years and many of them experienced the usual difficulties that young firms have of becoming steadily profitable. But their products have caught on quite well. Most hydroponic farmers say they can't meet demand, and the little industry seems to have successfully weathered a history of experimentation marred by promotional schemes and frequent failures. "This is the wave of the future," one grower exults.

It's really more like the wave of the past. Hydroponics is at least as old as the floating gardens of Xochimilco in Mexico, which date from the 16th Century at the latest. But it remained little more than a curiosity until recently, when research perfected new equipment

and techniques that make production commercially feasible. Now some sizable, well-financed companies are in
the business; Superior Oil Co., for example, is selling
hydroponic produce through Environmental Farms Inc.,
a Tucson-based subsidiary.

Hydroponics isn't big enough yet to scare field-crop
farmers. By far the biggest company in the game is Hydroculture Inc., a publicly owned concern in Glendale,
Ariz., near Phoenix, which sold four million pounds of
tomatoes in 1972 out of some 2.4 billion pounds consumed in the U.S. Furthermore, it's unlikely that hydroponics will ever squeeze out field crops; the tank-grown
product costs too much, and, though prices seem likely
to decline with growing production, they aren't expected to be competitive within the foreseeable future.

Still, many housewives seem willing to pay an average of about 20 cents a pound more for hydroponic tomatoes and even a higher markup for cucumbers. Safe-
way Stores and Jewel Foods, both major buyers of Hydroculture tomatoes, report climbing sales and "a real
following" for the tank-grown version.

Growers say that's because all hydroponics produce
is vine-ripened before picking, which is thought to add
to the flavor. Moreover, growing in the totally controlled environment of hydroponics eliminates the need
for pesticides, herbicides and other chemicals that are
repugnant to environmentally conscious buyers.

Some growers see a huge potential. The rising costs
of land and increasing pollution are making field farming less and less feasible, they argue. "I can foresee a
time when it will be impossible to grow a decent field
crop," says John Harsany, president of Hydrokist Inc., a
division of Luckey Laboratories Inc. of San Bernardino,
Calif.

Some of the unique characteristics of hydroponics
might make it an important process in arid lands, those

with severe climates, and those that suffer from land shortages. The United Nations Food and Agriculture Organization is studying it. Experimental units are operating in Kuwait, Lebanon, Israel and elsewhere, and Hydroculture has shipped a unit to Moscow. Malcolm Lentz, president, says: "Ultimately, the foreign potential is greater than the U.S. potential."

For one thing, Mr. Lentz says, "you can grow anything you can grow in soil, and you can do it on submarines, on the tops of buildings, or in space stations, if you want to." A single hydroponics unit costing $20,000 and covering an area only 26 feet by 128 feet is incredibly prolific; it can yield as much tomato poundage in a year as four to eight acres of land because it operates all year round. Hydroculture also raises grass for cattle feed in a "magic meadow" unit that grows a new crop every week; a growing area of only 1.4 square feet yields a ton of grass a year.

Climate control makes hydroponics units immune to harsh weather. In Anchorage, grower-distributor Mike Mosesian produces tons of tomatoes while snowstorms swirl outside. And in the arid, sweltering Middle East sheikdom of Abu Dhabi, a research team from the University of Arizona successfully raises squash, beets, cucumbers and kale. (Hydroponics is thought to have particular potential for arid areas, since it makes very efficient use of water; the plants use only what they need, and the rest is recycled).

For the moment, however, growers are more interested in meeting U.S. demand for their vegetables, which are marketed as gourmet items. Roger Marr, a grower in the Las Vegas area, says he could sell many times more tomatoes than he is able to raise with his five units, and he plans to double the number. Others are planning similar expansions. In 1973, Hydroculture landed a $2 million contract to build and manage a

100-unit complex in the Southeast for Trans Pac Leasing Inc., Portland, which planned to rent it to two other concerns. At the same time, the company researched the commercial growing of lettuce, cabbage, okra and strawberries, and also experimented with spices and pharmaceutical plants.

Mr. Lentz of Hydroculture says that in selling hydroponic units the company often must overcome the skepticism of farmers and other customers stung in the past by fly-by-night promoters selling gear that didn't work. Others never got off the drawing boards. One Arizona company got the idea of building its greenhouse units by spraying foam plastic over inflated balloons and then letting the air out; when that happened, the whole structure collapsed and so did the company.

And, it must be added, not everybody likes hydroponic produce. The San Diego Zoo has a "magic meadow" grass unit to feed some of its animals, and Dr. Lynn Griner, head of the Zoo's health department says: "The unit has been well accepted by the deer, monkeys, turtles and hippos, but all of a sudden our giant tortoise won't touch it."

—HAL LANCASTER

Flying Farmers

Early one spring morning, a white, red-trimmed biplane took off from a grassy field in eastern Kansas and began swooping back and forth over the flat river bottom land. Within six hours, two fields totaling 300 acres were seeded by air with hybrid corn and had received a dose of fertilizer to boot. The same job, if done by conventional methods, would have taken two men at least three days.

Thus, the farmer has strapped another harness on the air age. Not that this is an unusual development. For all their supposed conservatism, farmers were among the first airplane buyers more than 50 years ago, and hayfield landing strips have become increasingly common in rural America ever since.

"The plain fact is, an airplane is becoming nearly as essential as a tractor to a farmer or a horse to a rancher," says David Martens, president of International Flying Farmers, an association of about 9,000 airborne agriculturists in the U.S. and Canada. "What with high prices for equipment and labor—when you can find labor at all—and steady consolidation of farms," Mr. Martens says, "a farmer has to get more out of his land and each hour of his day in order to make a profit. And the airplane is helping him to do it."

Farmers, in fact, are using their planes for almost everything except milking cows and plowing. Apple growers in Washington hire helicopters to spread pollen in their orchards in the spring when winds or cool weather hamper the bees' activity. In the summer, corn farmers are buzzing their fields to stir up more pollen, resulting in "heavier, better-filled ears" of corn, according to Illinois farmer Raymond Roller.

Howard P. Wackman, a 30-year-old rancher near Sacramento, Calif., hunts coyotes from his plane to protect his 1,100 sheep. And Harold Clement, who has a big ranch in western Manitoba, herds about 300 horses from pasture to pasture with his two-seat Hughes helicopter. In the Western plains, where cattle graze on open range, airplanes are used to spray them for mosquitoes and other pests. "It would take weeks to round them all up for spraying," says one rancher.

Spraying or dusting crops with weed and insect killers was one of the first jobs of airplanes on farms: The first recorded crop-dusting venture took place near Lima, Ohio, in 1921. The practice is still growing, and today more farmers are spreading fertilizer from the skies, too.

But the technique gaining widest attention on farms these days is aerial planting. Southern farmers were the first, starting in the late 1950s. More than 90% of the nation's annual rice crop is planted by air, says a spokesman for the Rice Council, a trade group.

In recent years, however, aerial seeding has spread north into the Midwest farm belt. Wheat, soybeans, pasture grass and other crops increasingly are being seeded from above without a noticeable effect on yields or quality. In 1968, two Iowa farmers each planted 10-acre fields of corn by air in the first aerial seeding experiment with that grain, the country's biggest crop. The two men harvested yields ranging well above 100 bushels an acre in what turned out to be a disappointing year for corn. The practice has spread since then.

The two fields in eastern Kansas, owned by M. R. Hudson, a Kansas City industrialist, were seeded by air because of weather problems. "I've got a lot of my corn in already," Mr. Hudson said that morning, "but those fields were so darn wet from the heavy rains we've had that aerial planting seemed like the fastest way to get a crop in."

Besides saving planting time, aerial seeding requires a less thorough job of soil preparation. Once sown, the seed still must be covered with soil, so the farmer does have to pull equipment through the field, anyhow, though this process is much faster than if he were planting on the ground. There are some disadvantages, though. Unless the seed is evenly dispersed, the crop will grow in uneconomical clumps, and combines harvesting aerial-seeded corn need to travel a bit slower to bite their way through the jungle of stalks.

New as it is to the north, aerial-seeding has already helped boost farm output. Farmers in the mild-climate belt from Kentucky west to California, for example, are wringing two cash crops a year from their land, with the help of ample fertilizer and airplanes, where only one crop grew before.

These farmers grow a normal summer crop of, say, soybeans or corn. In late summer, before harvest, the fields are aerial-seeded with a small-grain crop, such as winter wheat, barley or rye; the seeds are nourished and sheltered by the falling foliage of the maturing crop. In the fall, the summer crop is harvested without any damage to the sprouting winter grain, which ripens and is harvested early the next spring before time comes to plant a new summer crop.

A farmer restricted to ground planting normally couldn't put in two crops, as he would have to wait until the summer crop was harvested, when it would usually be too cold and wet to get a new crop started before winter.

But agricultural aviation is provoking controversy, too, particularly because it involves application quickly over vast areas and is consequently more prone to error.

For instance, aircraft are widely used for spraying defoliants like 2,4,5,-T on power-line rights-of-way and in hilly areas where farmers want to clear out scrubby timber. Use of such defoliants, which the U.S. halted in Vietnam after tests on rats indicated they might cause cancer in humans, has aroused plenty of argument.

Agricultural airmen stress the importance of careful application, but concede there can be mistakes. "We work mainly in an agricultural area, and people realize the hazards and benefits," says W. G. Irwin, secretary-treasurer of American Dusting Co. of Chickasha, Okla. "We stay away from congested areas where someone might see a bird drop dead and get upset."

Chemicals that drift from one field to another are especially troublesome. Norman Akesson, a professor at the University of California at Davis, cites an instance where a herbicide being applied in one California field drifted to a crop of sunflowers, completely destroying them. E. R. (Butch) Kinnebrew, manager of crop-aviation underwriters for Crump, London Underwriters of Memphis, tells of a case in Tennessee where drifting pesticides wiped out $25,000 worth of catfish. Aviators carry insurance for such calamities.

But airmen argue that these drift problems can be prevented with more careful application, and avoidance of the dusts that drift most easily. Most materials are applied in liquid form, and chemical companies such as Dow Chemical Co. have developed compounds that solidify pesticides and herbicides into heavy droplets that aren't supposed to drift at all.

Airplanes are changing the way farmers do business, such as spurring cross-country farm purchases. For instance, a Missouri farmer, a South Dakota farmer-rancher and a Colorado rancher, all members of the flying farmers group, bought a 34,200-acre ranch in southeast Montana. "All three of us happened to be together one Saturday when we heard this place was for sale," says L. Karl McDonnell, the Missouri farmer. "Next day, we flew up there to look it over, and decided to buy it. We have men running it for us, but the management end is up to us—which wouldn't be possible without our planes."

Planes are being used more to keep track of farmers' activities, too. Rural bankers, for instance, are taking to the skies for aerial inspections of farms before making decisions on loans. Federal and state agriculture officials are making aerial inspections, too, to get a bird's-eye view of how crops are progressing.

Cost-conscious farmers also are flying farther in search of equipment, livestock and supplies. One Great Plains farmer says he regularly flies to Canada to look over what he thinks are bargains in machinery; one year in Saskatchewan he bought a combine he needed and saved enough, he says, to pay all his airplane expenses for a couple of years. Livestock sales are being attended by a greater number of ranchers and feeders from hundreds of miles away, auction concerns say. One rancher says he saves 50% on lumber for fences and farm buildings by flying his plane to another part of his state to do his buying.

When they're on these trips, though, farmers don't forget the world below them. They're always thinking, observing and calculating, as shown by the experience of Robert R. Ingle, who one day was flying from Ft. Madison, Iowa, to his 3,400-acre farm and ranch near Cavour, S.D.

"The winds were bad so I had to fly low," he recalls. "I started observing the cattle feedlots that are all over Iowa, and by counting the cattle and looking at the feed bunkers, I figured that the feedlots were all but full to capacity with cattle—which meant that prices for feeder cattle would likely be falling." Within four hours after landing at home, Mr. Ingle had phoned a buyer and contracted to sell a herd of his cattle at the current price. "Two days later, the price had dropped more than $2 a hundredweight," Mr. Ingle chuckles.

The airplane is slowly making farmers more cosmopolitan, too. Instead of being forever stuck out in the middle of nowhere, or at best heading into town on Saturday night, more farm families are taking jaunts of a few days each, whenever the work-load lightens, to the mountains or the beach.

The Sol M. Smith family, which farms near a small town in northern Oklahoma, borrows their four-passenger plane from farm chores frequently. "Many times a year we fly to Kansas City, Dallas and other cities to see plays and take in other events," says 43-year-old Mr. Smith. "A long stretch of life on the ranch can get pretty dull."

—JOHN A. PRESTBO AND
WILLIAM M. BULKELEY

Farming by Computer

Kendall Luce, a 31-year-old Madison, Maine, dairy farmer decided to increase his milking herd to 130 cows from 70 and to grow 250 acres of corn instead of 70.

Nothing very unusual perhaps, but what makes those decisions somewhat out of the ordinary is that they were suggested by a computer. Mr. Luce subscribes to a computerized farm planning service offered by Doane Agricultural Service Inc., headquartered in St. Louis. He fed the computer information on his basic resources such as land, capital, buildings and equipment. The computer analyzed the data and then offered a plan for putting the resources to work in the most profitable way.

Instances of computers calling the shots in farming are no longer uncommon. After years of being used by farm-credit bureaus and universities for such peripheral farm functions as accounting and record-keeping, computers are moving into the heart of farm operations, helping answer such central questions as what crops to raise, how many acres to plant and what equipment to buy.

Farm experts say there are two main reasons for the trend. A computer can supply more data in less time than a farmer alone can compile, and, the experts explain, the additional information is just what farmers need to combat a profit squeeze. "Most farmers who stay in business will be using something like this (computerized planning) within three to five years. There are too many benefits to be reaped not to use it," J. W. Hackamack, Doane chairman, said in 1972.

The Doane service was initiated in 1971 and in

about a year about 15 farmers with a total of some 30,-000 acres signed up. An attempt to market a similar service in the late 1950s was unsuccessful, mainly because the $5,000 cost to farmers was too high, the company says.

The cost of the new program isn't cheap, ranging up to $3,000. This is considerably more than most other computer services charge, but the Doane program also is more comprehensive than most others.

A farmer must answer 35 to 50 pages of questions about his operation—things like the amount of labor available in his area during various seasons, the type of machinery he has, how much capital he's willing to invest and even what sort of return on investment he's expecting. The answers are fed into the computer, which considers literally thousands of possibilities that an individual could never calculate by hand.

As a result, the computer is pretty hard to out-guess, Mr. Hackamack says. He cites an instance when it appeared obvious that corn ought to be the crop of choice, but "the computer threw it out because there wasn't enough labor available at critical times of the year. It recommended wheat, instead," he says.

A more modest computer service has been developed by Michigan State University in East Lansing. Called Teleplan, it uses a computer, named Synthia, which is programmed to answer 50 specific agricultural questions. Farmers in Michigan and neighboring states can query Synthia by using touch-tone telephones at offices of county extension agents. The computer responds in a mechanical monotone voice.

Since its inception in 1970, Teleplan has received more than 13,000 calls. One question, dealing with dairy feed rations, has been asked more than 5,000 times. Other questions, and answers, deal with topics such as the type of equipment a farmer should buy.

Computerized farm-planning services, especially those like Doane's, aren't for everyone. "A farm should have more than $40,000 in sales, at least 800 acres and three or four crop opportunities, as well as one or more livestock opportunities," Mr. Hackamack says. One Doane consultant recently recommended against a successful Louisiana cotton and soybean farmer buying the service because the farmer didn't want any part of raising hogs or chickens, thus eliminating possible alternatives and restricting the computer's potential usefulness.

Some farm experts express doubts about the value of a total planning package like the Doane service. Says one: "There are just too many variables."

Nonetheless, those who have subscribed seem satisfied, though it's too early in most cases to show concrete results. Mr. Luce, the Maine dairy farmer, believes the service will pay off. "I can already see how the changes will bring more profits," he said, and he estimated that within three years of making all the suggested changes in his operation, his net income should climb to $27,000 from $12,000 in 1971.

Another satisfied customer is Scott Reiss, a 25-year-old Plains, Kan., farmer who turned to Doane for guidance after taking over a 2,200-acre farm following his father's death. In May 1972, he added corn to the farm, planting 130 acres at the computer's suggestion. Many of his neighbors were growing alfalfa, but the computer selected corn over alfalfa because it would give Mr. Reiss the 14% pretax return on investment he had specified.

The computer not only can boost profits, but it also can minimize losses. Donald Lomen, Doane marketing director, tells of a tobacco, corn and soybean farmer who had incurred losses for the previous few years. "It was questionable whether he could make money under any circumstances because of the small size of his farm, but a computer analysis showed that by turning the farm into a tenant operation, he would have a chance to break even," Mr. Lomen says.

Sometimes farmers derive a side benefit from computerized planning by using the computer analyses to show bankers the soundness of lending money. "Bankers love to have something like this to support a loan application," Mr. Hackamack says.

—DAVID P. GARINO

More Muscle

Leon Miller figures he's one up on the big packinghouses that buy cattle from his Billings, Mont., feed lot.

His secret weapon: A teletypewriter that for up to 10 hours each day churns out information on prices, sales volume, weather conditions and other key aspects of the cattle business.

Mr. Miller subscribes to Cattle-Fax, a market-information service offered by the American National Cattlemen's Association. It cost him $260 a month in 1970, but he says it was easily paying for itself. For example, he once coaxed an additional $4,000 from a packinghouse buyer who had bid for 800 head. "After the buyer found I had the same information he had, he saw there wasn't any reason to keep bargaining," Mr. Miller recalls with a chuckle.

Mr. Miller and thousands of farmers like him are flexing some newly gained marketing muscle. They're relying increasingly on nationwide information services like Cattle-Fax. And they're banding together in large marketing cooperatives that negotiate prices and control the flow of produce to the marketplace.

Farming co-ops have been around for years, of course. But rarely have they been big enough or powerful enough to exert more than mild pressures on local food prices. Now things are changing, the experts say. Established farm groups are broadening their marketing activities, and new groups are sprouting. Some of their methods aren't new, but they've never been tried on the scale that farmers are attempting today.

"If these efforts are successful, there's simply no

doubt that prices for the products involved will rise," says a U.S. Agriculture Department official. Indeed, there was evidence throughout the early 1970s that the marketing groups were a factor in pushing up food prices.

The impact on the housewife's food budget could grow in the years ahead as the marketing groups become stronger. And farm authorities agree that the emergence of these groups signals that farmers are breaking with tradition.

"For years the emphasis in agriculture has been on increasing production," says William C. Helming, general manager of Denver-based Cattle-Fax. "Farmers were showing off their efficiency by producing stuff and then dumping it at whatever price they could get. Now they're realizing that they should be paying more attention to marketing, so they're adopting some of the techniques developed by industry."

One reason for this change is that the number of farmers has been declining rapidly, and those that are left control greater shares of the total production. "Fewer farmers, each with a big investment to protect, are more likely to band together for marketing purposes," J. Phil Campbell, U.S. Under Secretary of Agriculture, observed.

"It has taken a new generation to get the marketing power we need," says Robert Johnston Jr., a rancher with 110,000 acres near Pueblo, Colo. "Grandpa was an individualist who looked at ranching as a way of life. But, hell, we can't eat romance, and the only way we can get a decent price is to work together."

There's some skepticism, however, as to whether farmers, long known for their stubborn independence, can work together in marketing groups over the long run. "Once the prices go up greedy farmers start selling," says Roby L. Sloan, agricultural economist of the

Federal Reserve Bank in Chicago. "That's why withholding tactics—shooting pigs, dumping milk, burning potatoes and the like—have never worked."

Nonetheless, some consumer advocates, including Ralph Nader, are taking the farmers seriously. They've already suggested a Government investigation of the marketing innovations. Farm groups are permitted under the Capper-Volstead Act to influence prices of agricultural products, but the act forbids "undue influence" and doesn't shield the groups from antitrust actions.

"All that would have to happen is for just one of these groups to get a grip on the supplies of a single food and push the price way up," says a high-ranking Agriculture Department official. "Then all hell would break loose, and the Government would crack down on every farm marketing group."

Most marketing groups say they don't want to restrict supplies but only to keep them flowing just enough to satisfy demand and avoid a glut. Some groups do have a degree of supply control, though. United Egg Producers, for instance, has the power to divert up to 5% of its members' eggs from grocery shelves to manufacturing uses, such as mayonnaise and dried eggs for cake mixes. The purpose is to buoy up prices in the consumer market in a period of excess supplies, United Egg explains.

United Egg also gathers and disseminates national egg-marketing information in twice-weekly telephone conference calls to members. Using this data, the group recently began computing farm egg prices, which it recommends its members use as an asking price for their eggs.

Farmers say that keeping close tab on marketing statistics helps them avoid bringing their products to market at the same time that many other farmers also are selling. Such bunching of sales drives down prices temporarily; in this situation chain stores normally buy extra quantities of foods that they then offer as "specials."

Other marketing groups, like the fruit, vegetable and nut cooperatives on the West Coast and in Florida, and the big dairy cooperatives, are highly organized outfits that do everything from negotiating prices to controlling the movement of the commodity to market.

Besides negotiating prices, some groups are assuring themselves of markets as well. The National Farmers Organization and the American Farm Bureau Federation, for example, are active in contract farming, where members agree to provide a specified amount and quality of a commodity in exchange for a set price paid by packers or processors. In New Jersey, the Farm Bureau has negotiated a tomato contract with Campbell Soup Co. for several years. Campbell agrees in advance to pay a set amount for the tomatoes.

Hogs, broiler chickens, fruits and vegetables are among the foodstuffs that are sometimes marketed this way. So far, the method has worked best in limited areas where the farm group involved has a strong membership that produces a sizable share of the commodity under contract.

In the long run, marketing groups may boost farmers' income more by lowering marketing costs than by raising prices. United Egg, for instance, has a program of directing members' eggs from areas of excess supply to locations where local farmers can't meet demand, bypassing the egg dealers and handlers who traditionally have performed this function for a fee. "Not only do we cut out some middlemen, but we avoid sending eggs to places where they're not needed, causing the local price to fall," a United Egg spokesman said.

—JOHN A. PRESTBO

Part Three

VERSUS
MOTHER NATURE

For all the skill, technology and human effort that makes U.S. agriculture so productive, the weather still makes the difference between feast or famine. It was bad weather throughout much of the world that increased foreign demand for U.S. foodstuffs in 1972, 1973 and the first part of 1974—causing food and fiber prices to rise sharply. Nature can clobber crops in a variety of ways—too much rain or too little, too hot or too cold—and so can man-made pollution. But the result is always lower yields. "Never predict the size of your crop till it's in the bin," is the advice of many a farmer who has watched a year's work and profit ruined by a spell of bad weather. Unexpectedly good weather can cause problems, too.

141

Bitter Harvest

A succession of snowstorms, torrential rains and floods wiped out close to $1 billion of livestock and crops in the spring of 1973, and delayed farmers from planting their crops.

The adverse weather started in the fall of 1972, causing long delays in harvesting; in fact, some of the 1972 crops of cotton and soybeans were still in the fields at planting time the following spring. Farmers couldn't do their normal fall fertilizing and cultivation, which meant a double dose of work had to be done in the spring.

But that's when unusually wet and cold weather hit much of the farm belt, with these results:

—Fruit and vegetable crops were severely damaged. Onions, potatoes, peaches and apples were among the items that suddenly became scarce and costly.

—Thousands of cattle and calves were killed by snowstorms or by being bogged down in mud. This loss reduced the number of cattle coming to market.

—Wet fields were too soggy to do much plowing or planting until late in the spring.

Not all farmers suffered from bad weather. In the northern Great Plains, for instance, seeding of spring wheat, oats and other grains proceeded ahead of schedule. And the rain and cool weather helped make pastures especially lush. But for the most part, the weather was bad news for farmers and consumers alike.

"We probably had the worst winter with the greatest impact in history," said a spokesman for the American National Cattlemen's Association in Denver. From November 1972 through April 1973, the trade group estimated 250,000 head of mature cattle were killed by storms. That was the biggest numerical loss in about a century; it had unprecedented price impact in 1973 because each animal produces much more beef than 100 years ago and because beef prices were buoyed by strong demand.

In all, about 120 million pounds of beef were lost worth $156 million at retail and amounting to about half a week's national beef production. In addition, many cows and calves were lost. It would take ranchers a couple of years to replace these animals.

The spokesman for the cattlemen's association said he counted 34 storms in the Denver area from November to April, sharply above the 10 to 12 storms during a normal winter. Some of them were extraordinarily severe, which exacerbated livestock losses.

On April 8 and 9, for instance, Iowa was hit by a storm with winds of up to 60 miles an hour that dropped nine inches of snow over most of the state and up to 19 inches in some places. The storm killed 365,000 livestock animals, including 107,000 cows, calves and steers; 214,000 turkeys, most of them eight to 10 weeks old; and 44,000 hogs, sows and pigs. Some experts said it was the biggest livestock loss from a single storm.

Another kind of bad weather—unseasonable spring cold snaps—crippled production of many fresh fruits and vegetables. For example, Georgia estimates that it lost as much as half its $18 million peach crop. Most of these peaches were to be sold as fresh produce.

In some instances, the cold snaps followed periods of abnormally warm weather, and the reversal proved too much for the fruit crops. In Indianapolis, the temperature for the first 15 days of March averaged 54 degrees, which was 17 degrees above normal; but the first 15 days of April averaged only 44 degrees, four degrees below normal, which included some freezing weather.

As a result, Indiana lost 75% of its peach crop and 20% to 25% of its apples.

Cold weather cut the Southern Illinois peach crop by 85% and the state's apple crop by about 50%, said James West, an official of the Illinois Fruit Growers Association. An official of the Michigan Fruit Canners, a trade group, estimated that as much as 70% of the state's sour-cherry crop was lost to freezing temperatures. In North Carolina, some damage was reported to apples, peaches and strawberries.

Some fresh vegetables were hit hard, too. The only onions that are marketed in early spring come from Texas. But rains there delayed harvesting up to three weeks, and the crop was damaged, too. Onion prices spurted to record levels, and in some areas there weren't any onions available at all. Growers in the Rio Grande Valley were getting a record 30 cents a pound—a sixfold increase from a year earlier—and in Chicago, the wholesale price rose 30% from the previous year. Potato prices jumped because of bad weather, too.

Even more worrisome than damage was the possibility in April that continued bad weather would delay plantings. The best time for corn plantings to get the highest yields is in the first half of May. The longer the rain continued, the fewer acres would be planted in corn and the lower-yielding they would be.

On the other hand, several days of drying weather would help greatly because modern equipment could perform the plowing and planting chores more quickly than in previous years. And ample soil moisture would get the crop off to a good start.

"Farmers around here can't get their tractors out of the sheds, let alone do anything in the field," said Dale Millis, an agronomist in Southern Illinois, in late April. "The soil is 100% water-saturated, and it will take two weeks of dry weather before spring tillage can begin, and even then many farmers will be working wet ground."

In Iowa, the top-producing corn state, plowing for the 1973 crop was only 35% completed in late April, compared with 74% at that time in 1972; the average for that time of year is 71%. In Illinois, the second-ranking corn producer, plowing was also 35% completed, compared with more than 75% a year earlier.

As of early April, more than 15 million acres of farmland were under water. Much of this land was still flooded a month later, and the rest was so supersaturated that it would take another month to dry out. Some of it wasn't planted in 1973 at all.

Cotton was especially hard-hit by flooding. Some 500,000 acres of Mississippi River delta land, where most of the high-producing commercial farms are located, was under several feet of water in late April. In the state of Mississippi, "in the past six months, we've received a normal year's rainfall," said Mark Freeman, executive assistant to the commissioner of agriculture.

The state had expected the planting of 1.6 million acres of cotton in 1973, Mr. Freeman said, but only 0.5% had been planted by late April, and two-thirds of the land wasn't even prepared for planting. At this time of spring, a third to a half of the land is normally planted. The prospect of weather-reduced cotton plantings, along with strong export demand, pushed cotton prices sharply higher.

In Mississippi, Tennessee and other Southern states, some of the 1972 cotton and soybean crops hadn't been harvested by spring. Because of the prolonged wet weather, "we'll be in the position of planting for 1973 while we're still harvesting for 1972," said John Allen, an official of the Agricultural Stabilization and Conservation Service in Nashville.

—NORMAN H. FISCHER

The Summer of Agnes

When the Susquehanna River receded in June 1972 after the devastating floods in eastern Pennsylvania, it took with it 61 prime acres of topsoil from Donald Barth's farm in the little town of Mehoopany.

The flooding, caused by Tropical Storm Agnes and the heavy rains that followed it, also resulted in enormous crop damage. Some 350 acres that had been planted in corn and alfalfa produced next to nothing. The result: Rather than buy feed at inflated prices for his 160 head of dairy cattle, Mr. Barth considered selling his herd.

Thousands of farmers throughout the East suffered similar severe crop losses. And, as usual, when farmers are pinched, others feel the pain. Eastern food processors, produce dealers, and housewives shopping for fresh fruit and vegetables saw the impact in shortages and, in some cases, drastically higher prices.

While public attention generally was focused on the huge destruction of homes and business property in Wilkes-Barre and other more populated areas, Pennsylvania Gov. Milton J. Shapp declared that "in the final analysis, damage to our farmland may result in the most serious long-range problem for our state."

The governor figured the loss in crops, livestock, farm buildings, equipment and soil erosion at "a staggering $300 million." In New York, the agricultural loss was put at $100 million. And Maryland officials said farm damage was about $24 million. In all, six states had total agricultural losses estimated to exceed $450 million, making "the Agnes summer" one of the worst U.S. farm disasters on record.

Nearly every crop was affected. In New York State, the onion harvest was 42% below normal, 53% of the bean crop was lost and severe production declines were evident for lettuce, cabbage and other produce. Vineyards in Pennsylvania and New York produced markedly fewer grapes.

By far the hardest hit, however, were Pennsylvania's dairy farms—the state's most important agricultural business. In many cases grazing lands were flooded, forcing farmers to feed their cattle on the oats, rye, barley and corn they had planned to harvest for use in the winter or for sale. What little hay and grain the farmers were able to bring in was often deficient in nutrients because they grew overripe in the marshlike fields that defied the use of farm equipment. The cost of replacement feed soared. By late August 1972, hay cost $80 a ton, up $12 to $15, according to Mr. Barth, the Mehoopany dairy farmer. Other farmers paid up to $92 a ton for hay and similarly high prices for soybeans shipped in from Indiana and Illinois.

Furthermore, the dairy farmers had to absorb such increases themselves. The Pennsylvania milk-marketing board fixes the price farmers get for their milk, and there isn't any way to pass on added costs. Gov. Shapp, however, sought an increase in the farm-milk price to help ease the farmers' plight.

Canners and other food processors also reported difficulties stemming from crop shortages. One Eastern company, faced with contract commitments for green beans and peas, was forced to buy finished products from a canner in another region and relabel them. Another, Hanover Brands Inc., among the largest food processors in Pennsylvania, said it suffered damage "that you can't exaggerate" to fields it normally draws from. The company said it brought in green beans from as far away as Wisconsin and Michigan for canning.

Hanover's vice president for agricultural and technical services, Nasmy Elehwany, said the company's contracted farmers lost their entire planting of sweet peas and about 50% of their corn yield, all in a year that looked "quite good" before the heavy rains.

Other companies operating in the Northeast reported similar problems with the tomato and cherry crops, for example. Notably, however, two big food processors with Pennsylvania operations, H. J. Heinz Co. and Campbell Soup Co., reported no crop-procurement problems at all, chiefly because of extensive sources outside the Northeast.

Although the canners, in order to remain competitive, didn't pass along the higher costs as increased prices, prices of fresh fruits and vegetables rose because of the storm. Produce dealers who relied on truck-farm crops grown in the Northeast generally turned to the West and Midwest for supplies—at higher cost to them and to their customers.

Charles Greco, owner of Norristown Wholesale Inc., outside Philadelphia, said fresh tomatoes sold in late August at about $6 for 30 pounds, up $1 from a year earlier, at his outlets in several states.

Some 44 Pennsylvania counties and two in New York were approved for federal low-cost grain under the emergency livestock-feed program, the U.S. Agriculture Department said. Loans at low interest rates also were made available to farmers.

However, the agricultural disaster became "a political football," as one farmer calls it. Democratic officials in Pennsylvania charged the Agriculture Department in Washington with foot shuffling and "a poor attitude" toward aiding farmers. The Agriculture Department denied this charge, countering that it did "everything legally possible." A spokesman said: "We feel like we've been on top of the situation and that we had money into these (disaster) counties before the water was off the land."

Specifically, Pennsylvania Agriculture Secretary James A. McHale contended that the federal department rejected the state's request for "herd-preservation" assistance; this would have provided temporary free feed for the hardest-hit farmers in six Pennsylvania counties. Instead, the Agriculture Department chose the emergency feed program. A department spokesman responded that the free-feed program, which lasts up to 30 days, is instituted only in the most extreme situations "where there are no visible means of supporting the animals." The spokesman terms the low-cost grain program generally "more beneficial" to the area because it doesn't expire until the emergency ends. Taking a potshot at Secretary McHale, he added: "I think he would rather assault (Agriculture) Secretary (Earl) Butz than to talk to us about getting the job done."

Some farmers, at least, had grown accustomed to the political disputes as well as to what they considered the somewhat plodding progress of federal assistance. And they looked at the existing aid programs as small indeed compared with the massive extent of the damage.

One Pennsylvania farmer said: "There isn't any program to satisfy people who lost fields. We can't reproduce topsoil—and much of it is lost for years to come. Farming is always a gamble, and all farmers know this. But a disaster like this isn't a part of the gamble that anybody can expect."

—ROY J. HARRIS JR.

Praying for Rain

The stores ringing the courthouse square of Altus, Okla., weren't crowded. On a Saturday afternoon—normally a busy time—only a few women shoppers were seen browsing among the dresses in the Powder Puff shop. And down the street at McPherson's Ford, there wasn't a prospective car-buyer in sight.

For a drought of nearly two years' duration (1970-71) had hit this once-prosperous area with a vengeance. While droughts and abnormal dryness were experienced in many areas of the country during this period, including most states west of the Mississippi River, it was drier around Altus than it was in the Dust Bowl days of the 1930s. And the local economy has been severely jolted.

Some Jackson County farmers went bankrupt. Many others borrowed to the hilt, and the Pioneer Farm Center, owned by 11 farmers didn't sell very much seed or fertilizer. Neither did fertilizer dealer Robert Kerr, who spent most of this time heading meetings of the Jackson County Drought-Disaster Committee, which sought federal help for this ailing land.

This is bountiful cotton, cattle and wheat country —when it rains. But the showers that came to these parched plains were too little and too late, and even the gravediggers complained because the ground was so hard.

It was too dry for farmers to plant cotton. They sold most of their cattle because the pastures had shriveled. And the 1971 winter-wheat crop was ruined. "If it rains all day and all night for the next 30 days I still won't make enough wheat off my 2,000 acres to get the

seed to plant next year's crop," said farmer Earl Aber-
nathy.

Businessmen, cattlemen and farmers in western
Oklahoma, western Texas and eastern New Mexico were
the big losers, but the repercussions of the drought
reached beyond the boundaries of these states. Consumers throughout the country paid higher commodity
prices because of the drought. The price of cotton shot
up, too.

But prices dropped for the products of the
drought-hit farmers and ranchers. Early onions were
marketed out of the Texas Rio Grande Valley, shriveled
by lack of rainfall, brought only $1.50 a bag in April
1971, down from $2.50 to $3.50 a bag a year earlier.

Whenever the drought may end, the region will
need a long time to rebuild. At San Angelo, Texas,
where five inches of rain fell in mid-April 1971, rancher
G. C. Magruder Jr. said that in a month his 30,000-acre
spread might have enough grass for some of his cattle
and sheep to begin grazing again. But even with continued favorable conditions, Mr. Magruder didn't expect
his ranch land to return to normal for at least a year.
(The drought didn't break until early autumn of that
year.)

Texas Agriculture Commissioner John C. White
said: "It takes time for grass to replenish itself. It will
take several rains over a period of time." Mr. White estimated that it would take six weeks of favorable weather
before the ranges became suitable for grazing.

In April, cattle prices were somewhat depressed by
the premature movement of herds to market. According
to Mr. White, the sale of cattle at auction in San Anto-
nio in the first week of April was up more than 100%
from the like period the previous year. Cattle sales in
the Houston area were up nearly 100%, he said.

The drought was also reflected in the thinness of

the cattle moving to market. Jack Drake, manager of Producers Livestock Auction Co. in San Angelo, said some animals "look like they've been on a two- or three-week starvation. Some animals are so weak that they just fall to the ground when they walk off the truck."

Hayden Ellis, a 53-year-old, ruddy-cheeked rancher, is one of those with undernourished animals. Surveying a herd of shorn, mangy, bone-thin sheep, he sighs, "I don't know what to do. They're starving to death, and I don't know what I can do about it. It's been years since I've had sheep as thin as these."

While cattle ranchers worried about their animals, other farmers worried about their crops. Mr. White said in April that wheat in some areas of the state was "already too far gone to be helped" and cotton in the lower part of the state would certainly be affected.

Mr. White said that by April 1971, the drought had cost the state more than $400 million and that if it continued—as it did—"we can expect deterioration in farm-related industries—farm machinery, processing and banking—as well as in general business."

Texas welfare authorities alerted county judges to prepare for a deluge of requests from migrant workers for surplus commodities. Many jobs the migrants normally pick up on farms and ranches on their annual trek northward were eliminated by the drought.

The hardest-hit part of Oklahoma was the southwestern sector around Altus. "We need 10 inches of slow, soaking rains within the next 60 days to break this drought, and even if we get that much we're in for a real rough year," said Elmer A. Provence, an area farm-management agent for Oklahoma State University's extension service, in April.

In a good year, Jackson County receives 24 inches to 30 inches of rain. Within the 18 months through mid-April 1971, large parts of the county received less than 12 inches—the driest since the weather bureau began keeping records in 1913.

As a result, farmers failed to produce much cotton in 1970 and 50-mile-an-hour winds ripped top soil off the idle fields and ravaged pastures. In many places, farmers ploughed 12 inch-deep furrows to act as ridges against the wind, but fences were nevertheless being covered by drifting soil.

"The only reason we haven't got another Dust Bowl is that we've got large equipment to chew up the ground and hold it better than the farmers could in the 1930s," said Gary Jones, who farms land near Altus that has been in his family for generations. Mr. Jones added, however, that "all the farmers around here are losing money, and even if it starts raining tomorrow it will take us two or three years to get even and overcome our losses."

At least 25 farmers who have borrowed money from one southwestern Oklahoma financing agency were said to be on the brink of bankruptcy and many others were approaching that point with each rainless day. One Jackson County farm, purchased in 1968 for $500 an acre, brought $375 an acre at a distress sale in April 1971.

In adjoining Tillman County, where drought-ravaged wheat also was being attacked by green bugs, the banks had $6 million in outstanding loans to farmers. Many loans fall due during the summer, but there wasn't much in revenues from winter wheat normally harvested in May—and the farmers normally use these revenues to make their loan payments. "We'll hang on with them as long as there's room to breathe," said Frederick banker, C. M. Crawford.

—JAMES TANNER AND ERIC MORGENTHALER

Unwelcome Bounty

There was bad news on the farm: The harvest promised to be the largest in history.

That may sound like a paradox, but it wasn't. In 1971, the harvests of such major commodities as corn, wheat and grain sorghum were so abundant that they drove farmers' prices far below expectations. The resulting impact on farm income forced more farmers out of business and weakened the economies of rural communities across the nation.

The big harvest wasn't bad news for everyone by any means. Much corn, wheat and sorghum is used as animal feed, and their lower prices prompted breeders to increase their flocks and herds. This, in turn, meant lower consumer prices for meats and other animal products. (For a while, that is.)

In addition, even some farmers who hauled in record crops benefitted. That was particularly true of soybean growers. Despite the fact that the 1971 harvest was a record, the price of soybeans had risen about 7% higher than a year earlier during harvest time. The increase reflected the growing world demand for the versatile vegetable as a high-protein food supplement.

Many other farmers, however, saw mostly troubles in the huge quantities of grain they brought in from their fields. One reason for their huge harvests was the almost total absence of Southern corn leaf blight, a disease that destroyed some 15% of the nation's corn crop in 1970.

The spores that caused the blight survived the winter and were expected to strike again in 1971, so at the recommendation of the U.S. Department of Agriculture corn farmers sharply increased their plantings to compensate for the looked-for losses. Growers of wheat and sorghum did likewise, expecting that their commodities would be in strong demand if corn failed again.

But instead, growing conditions in the summer of 1971 were nearly perfect all across the country. Spring came early, allowing farmers to plant almost two weeks earlier than usual. There was plenty of rain in May and June, getting the young plants off to a good start. Southern corn leaf blight thrives in hot, humid weather, but the weather in the crucial July and August growing period was unseasonably cool in most places. The blight didn't survive; the corn did.

By Thanksgiving of 1971, the horn of plenty had overflowed in the southern Iowa community of Albia and spilled almost 50,000 bushels of corn into the streets.

On old U.S. Highway 34, entering the town from the west, was a mound of corn three city blocks long and five feet high—a total of 30,000 bushels. On Avenue C West, near the Goode Seed & Feed Co. elevator, there was a 10,000-bushel pile. Several thousand bushels more were heaped on North 3rd Street near the city park. Small boys ran through the corn barefoot and threw handfuls of it at each other.

"It's a monkey on our back," said Ralph Goode, owner of the town's only elevator. "We were facing a hard choice: Either find someplace to put all the corn, or start turning farmers away."

The latter choice wasn't popular with merchants in this county-seat community of 4,100 people. Like most farm-belt towns, Albia depends on the income generated by local farmers for its economic survival. Thus, when Mr. Goode's bins were filled to their 45,000-bushel capacity in October 1971, the city council was quick to grant him permission to store excess corn in the streets until he could build some more bins to store the excess.

Piling grain in the streets is nothing new in the western Great Plains, where wheat frequently is dumped on paved thoroughfares to await rail or truck shipment to market. But in the Corn Belt, it's highly unusual to have more corn than anybody knows what to do with, and long-time residents of Albia and other Midwestern towns with corn in their streets said they had never seen anything like the 1971 glut.

Even more corn was piled in farmyards and in vacant lots along railroad sidings throughout the Midwest. One such pile in Cruger, Ill., contained 250,000 bushels and covered an area as big as a football field. In Long Point, Ill., a corncrib burst from overloading, spewing 175,000 bushels on the ground.

Such was the aftermath of the record corn crop produced by the nation's farmers in 1971—some 5.55 billion bushels, 18% above the 1967 record and about 800 million bushels more than projected consumption. In Monroe County, Iowa, where Albia is, the corn harvest exceeded three million bushels, compared with 2.2 million bushels in 1970.

About half of the corn produced in the U.S. is fed to animals on the same farm where it is grown, and even more is used by neighboring farms. Consequently, it's normally advantageous for farmers to store their corn as close to home as possible, either in their own bins or in a nearby elevator.

In 1971, there was further incentive for keeping the corn close to the farm. The over-supply of corn pushed market prices to their lowest levels in 11 years, and many farmers were unwilling to part with their crop for the depressed returns. In late November 1971, the cash price for a key grade of corn in Chicago dropped to $1 a bushel, compared with $1.62 a bushel in 'June. Prices paid by elevator operators like Mr. Goode were even lower, because they didn't reflect the cost of shipping the grain to Chicago; Mr. Goode was paying 95 cents a bushel for corn.

But very few farmers were bringing corn to Mr. Goode for sale. Instead, they paid him to store it in hopes that prices would rise enough in the next months to cover the storage costs and allow some profit.

Many farmers in other areas wanted to hold their crops off the market, too, but the effect of the big harvest on storage space made this difficult. For instance, in McLean County in Central Illinois, which claimed to be the most fertile corn-growing region in the nation, there was space to store only about 30 million bushels of a 50-million bushel crop. "We'd planned to build more bins this year, but we didn't because we expected a small crop," said Paul Anderson, president of Hansen Winkle Grain Co., a storage firm in Bloomington, the McLean County seat. "Now it looks like a lot of corn here will be stored on the ground." Corn stored on the ground for even a week or so commonly deteriorates and loses part of its market value.

All this posed a cruel dilemma for farmers. "The situation now gives me a sinking feeling—I'm stuck if I sell or if I hold," said Ivan Britt, who farmed 200 acres in McLean County. He said that his fields yielded 134 bushels of corn an acre compared to just 92 bushels in 1970, but he added that "it didn't pay to have such a good crop. I'd rather get 100 bushels an acre and get something for it on the market. It was bad last year, seeing the blight kill my corn, but this isn't much better."

Harvests of other crops were equally bountiful. In all, total U.S. output of food and fiber in 1971 topped 1970 by 12%, the biggest one-year jump in a dozen years.

Wheat farmers, who comprised about 25% of the farming population, weren't as bad off as corn growers. Instead of the sizable price increase they had looked for in the fall, the Chicago quote in late October of $1.55 a bushel for their crop was down from the June 1971 high of $1.67.

Sorghum farmers likewise felt the big-harvest pinch. October prices of around $1.85 a hundred pounds for the grain are down from $2 a hundredweight a year before and $2.65 earlier in 1971.

The income of wheat and sorghum farmers wasn't off this year, but it didn't rise much either. "And with retail prices going up the way they have, this amounts to a loss of real income," an official of one farmers' group pointed out.

—FRANCIS L. PARTSCH

Take the largest wheat crop in history, combine it with the lowest grain prices farmers have seen since the early 1940s, and what do you get? Mainly, a farmer who isn't buying many refrigerators and color-television sets.

Except at Bismarck, N.D., where Chick Hale is doing what might be considered, for a wheat-farming area short on cash, a land-office business. A newspaper ad for his appliance store's fall sale read: "Color 25-inch console TV now only 440 bushels."

This return to the old barter system struck an immediate, responsive chord in this largely rural area. "I'm pulling in farmers from as far as 100 miles away," Mr. Hale said, "selling merchandise to people I never would have gotten in my store, otherwise." He estimated his customer traffic rose "a minimum of 10%" from the levels he had anticipated with the sale.

One of his customers traded 586 bushels of wheat for a refrigerator, range, dishwasher and dryer. All together he took in about 1,500 bushels on barter transactions in the fall of 1971. Farmers deposited their grain at a local grain elevator and turned over to Mr. Hale certificates covering the number of bushels traded.

Mr. Hale offered customers $1.50 a bushel for the wheat, which was valued at about $1.20 a bushel on the local market. Mr. Hale was betting that the price would rise to at least $1.25 a bushel, which would give him "a small margin." But in any case, "my primary objective was to entice the farmers into the store to get acquainted," Mr. Hale said.

—FRANCIS L. PARTSCH

Choked Crops

Around midday, a massive bank of brownish smog moved in from Los Angeles and lodged against the San Bernardino Mountains, hovering directly over the thriving romaine lettuce patch of the Yamano brothers, the biggest truck farmers around Corona, Calif. By the end of the day, the Yamano brothers no longer had a thriving romaine patch. They had a romaine patch that looked as if it had been burned with a torch.

Incidents like this have occurred with growing frequency in the past 20 years as air pollution has thrown a smothering blanket over rich agricultural areas from California to New Jersey. Dirty air has destroyed tobacco leaf in Virginia and potato plants in Michigan, stunted citrus yields in Florida and driven out truck farming in areas of Illinois, Pennsylvania and New York.

In the Los Angeles Basin, the agricultural and horticultural topography has been virtually transformed by air pollution. Most of the cut-flower industry has fled north; farmers like the Yamanos have abandoned efforts to grow leafy vegetables, and all growers have learned to accept often-severe damage to their other crops, such as citrus, alfalfa, barley, radishes, green onions, celery and tomatoes.

Estimates of the damage caused U.S. agriculture by air pollution vary widely, partly because some studies ignore losses that others include, and partly because all of them are based on guesswork rather than hard data.

In its 1970 annual report, the federal Council on Environmental Quality estimated that air pollution was causing at least $500 million yearly in damages to crops and livestock. A more conservative study by the Stanford Research Institute, matching pollution counts across the U.S. with the "known sensitivity" of crops to pollution, resulted in an estimate that dirty air costs growers about $130 million in direct injury to crops and ornamental plants yearly, based on 1964 data. (The study was financed by the Automotive Manufacturers Association, the American Petroleum Institute and the federal Environmental Protection Agency.)

A Stanford Research Institute plant pathologist, Harris Benedict, sniffs at the way some governmental agencies arrive at their higher loss estimates. "The latest estimate I've seen from the U.S. Department of Agriculture is $350 million just to crops alone," he says. "That figure was arrived at by splitting the difference between one expert who said $200 million and another who held out for $500 million. I'm serious."

Other plant pathologists and government officials, in turn, question studies like the Stanford Research Institute's. Air pollutants—mainly ozone, PAN (peroxyacetyl nitrate) and sulfur dioxide—damage plants in two ways. They directly injure the exterior of the plant, thus ruining its market value, or they act to suppress its water intake and thus stunt growth.

The Stanford group's study ignores the growth-suppression factor, which causes substantial losses in almost all crops, especially fruit and root plants (often without accompanying exterior injury). The study also does not encompass important horticultural industries like home plantings and flowers, or the ripple effects of air pollution, such as erosion after plants die and the cost of grower relocation.

Only one or two studies into growth suppression have been completed, but the findings thus far are startling. In one project, conducted by Ray Thompson, a plant pathologist at the Statewide Air Pollution Research Center at the University of California in Riverside, navel orange and lemon trees and zinfandel grapevines were placed in greenhouses in which pollutants were filtered from the incoming air. Mr. Thompson says the results indicate that dirty air reduced the poundage yield of the grapes by as much as 60%, navel oranges by 50% and lemons by 30%.

A similar study of tobacco leaf at the U.S. Agricultural Station at Beltsville, Md., showed growth suppression reduced leaf yields by roughly 20% to 40%.

Jim Yamano, of the Corona truck-farming family, doesn't need scientific studies to tot up his losses. They are almost entirely from direct, visible injury that leaves crops unmarketable. He can see the brown tips of his green onions, the mottling of his celery and the wilting of his romaine lettuce.

It doesn't take much pollution to ruin the market value of one of his crops. Exposure for as little as four hours to concentrations of oxidants as low as .1 part per billion—far below the level of irritation to humans—produces plant injury, the Statewide Air Pollution Research Center says. In these modest amounts, the two oxidants, ozone and PAN, leave plants mottled or glazed; sulfur dioxide bleaches leaf tissue.

But when air-pollution counts are high early on a hot, sunny day, with irrigation moisture in the fields, the losses can be total. "In the morning, the romaine will be beautiful, green and tender," says Mr. Yamano. "Then the smog comes, with moisture and heat and no circulation, and it's pitiful, like you burned each leaf with a match." Every time that happens to one of the Yamanos' 100-acre patches, it costs them $25,000 to $30,000 in unrecoverable expenses, not counting profit loss.

Under a federal Environmental Protection Agency program, county farm agents are being trained throughout the country to systematically tally and report visible, direct pollution injury such as the Yamanos' crops suffered. Sometime in the future, economists hope to plug loss estimates from both direct injury and growth suppression, plus all tributary effects, into a mathematical model that would accurately show what pollution damage is costing U.S. agriculture and horticulture.

Some skeptics ask, however, what good that kind of information would do, if ever obtained. Not much can be done about pollution damage to plants until stringent controls drastically reduce auto and factory emissions, a day that is decades away.

Some progress has been made in developing more smog-resistant strains in crops like tobacco, and in a couple of experimental projects crops have been sprayed with ascorbic acid, an antioxidant agent, to see if it would protect the plants from burn. But the sprays have proved largely ineffective, and developing resistant strains is, to some researchers, a painfully piecemeal and even wrongheaded approach.

"Looking for resistant varieties is really a miserable way to try to solve the problem," says J. B. Mudd, a biochemist at the Riverside center. "We're looking at the wrong end of the stick and assuming a very defeatist attitude—that we're going to have pollution forever."

Statistics published by the Los Angeles County Air Pollution Control District indicate that dirty air may continue for a long time, but it isn't getting much worse in the Los Angeles Basin. Since 1965, little change has occurred in the tonnage of daily emissions of oxidants, particulates and nitrogen oxides, although sulfur dioxide emissions have been cut in half.

However, you can't convince farmers with fields outside metropolitan areas that the situation is static.

"When we came here 20 years ago, there was maybe a week of damaging smog yearly," says Jim Yamano, whose family was driven out of the suburban San Fernando Valley, some 60 miles to the northwest, by stifling air pollution there. "Now we have 300-odd days of bad smog, and it's getting stronger and stronger. Growing produce in this area anymore is now just one big gamble."

—G. CHRISTIAN HILL

Part Four

FARMERS AND POLITICS

Agriculture and politics have been intimately entwined ever since—and even before—this country was founded. The political power of agriculture has diminished with the growth of urbanization, but it has by no means disappeared. In 1974, for example, the former head of a giant dairy cooperative pleaded guilty to bribing a government official to help bring about an increase in milk price-support payments a few years earlier. That same year, livestock ranchers persuaded Congress to grant them low-interest loans because they were being squeezed severely by climbing costs and plunging prices. The relationship of government and agriculture may be in a state of flux, but it is never likely to be severed.

Losing Clout

California's rugged coastline, majestic mountains, comfortable climate and lush valleys continue to lure Americans from near and far. But the more of them that come, the more pressure they put on the state's agriculture to move over and make room. There was a day when agriculture was so powerful it didn't need to budge. Now it has no choice.

The forces of urbanization and agriculture are clashing throughout the country, but the confrontation is of especially epic proportions in California, the nation's number one farm state. California's rich farmland—amounting to 36% of its 100 million acres—yearly yields more than 200 commercial crops. That's the most diverse output in the U.S., and it provides 25% of the nation's table foods and 40% of its fruits and vegetables.

For the past 25 years, California has led the nation in cash farm receipts. In 1972, the receipts were a record $5.1 billion. In 1973, they were nearly $6 billion. Agriculture's ripple effect in the state, according to agricultural economists, means another $12 billion to $15 billion in revenue for agribusiness industries. In all, agriculture accounts for a sizable chunk of the state's economy.

Even so, agriculture hasn't been able to withstand the effects of burgeoning urbanization, especially the fact that California's population tripled from 1940 to 1973 to about 21 million. Agriculture no longer can claim its former role as the dominant political force in the state. It has become just another special interest—

one that still wields considerable power, but not with the assured success of 25 or 30 years ago.

"Agriculture's political clout has been eroded fairly consistently over the past five or 10 years," says Richard Wilson, a California cattle rancher and chairman of the Planning and Conservation League, an environmental group. "Today, agriculture has to go around looking for coalitions in support of its position, whatever it might be." In fact, over the past decade the state's farming interests haven't always agreed even among themselves —sometimes splitting along regional lines or according to the size of farming operations on a variety of issues such as water distribution, land use and taxation, farm labor and pollution and pesticides control.

Agriculture's power in California began waning at the end of World War II, but the decline accelerated in 1966 when the state senate was reapportioned along the one-man, one-vote ruling of the U.S. Supreme Court. Before reapportionment, for example, Los Angeles County had only one state senator; now it has 15 of the state's 40 senators. Los Angeles' gain was agriculture's loss, as many rural counties lost their senate voice.

"Reapportionment has definitely meant a lack of interest in farmers' problems," says Richard Johnsen, executive vice president of the Agricultural Council of California, a trade group that represents California's farming cooperatives. "We're seeing the passage of legislation that is restrictive on agriculture without anyone taking into account the importance of agriculture in the state."

For instance, the state legislature in 1972 added to frustrations of large Southern California farmers by preserving the wild state of several north-coast rivers. Farmers and others argued that development of these rivers, particularly the Eel, would be necessary to replenish the state's water supply in future years. The lawmakers placed a 12-year moratorium on any development of the Eel.

Farm-group leaders believe the switch to an urban-oriented state legislature has also hurt agriculture's position regarding the emotional farm-labor unionization issue, which has bloodied California's vineyards and lettuce fields off and on since 1965.

After an early recalcitrance—and myriad violent incidents—California's agriculture industry has in recent years sought legislation, thus far unsuccessfully, that would establish election procedures for farm workers to choose which union, if any, they want.

On that, the farmers aren't far apart from their archenemy, Cesar Chavez and his United Farm Workers Union. The UFW, which in 1971 organized grape workers but which suffered some setbacks in 1973, favors worker elections. But most proposed legislation advocated by agriculture usually would ban the secondary boycott, a favorite Chavez tactic.

Some farm-group leaders chastise urban legislators for not passing farm-labor legislation. "The urban legislator is more likely to interpret the farm-labor issue in terms of individual human needs than from the standpoint of the agricultural industry or consider the contributions that agriculture has made," says Don Curlee, executive assistant of the Council of California Growers.

Galloping California urbanization has other effects, too. Some 22,000 to 64,000 acres a year have been taken over in the past two decades for housing tracts, shopping centers and highways. The value of prime farmland adjacent to sprawling urban centers has skyrocketed, thereby lifting property taxes to the point where some farmers can't afford them. Many have sold their land to bigger farmers or to developers.

To protect prime farmland, the legislature in 1965 passed a law that cities and counties may agree to tax farmland at a lower rate if the owners pledge to use it only for farming for 10 years. About 11.4 million acres have been preserved under this law, 2.7 million of which are considered prime farmland.

But the law "isn't working well at all in protecting urban fringe land," says Don Gralnek, land-use consultant to the state assembly. Some critics say large landowners have used the law as a tax shelter, which state officials acknowledged in saying there are plans to tighten up the law's provisions.

While some choice land is being lost, some less-choice land is being brought into production by huge irrigation projects. Authorities estimate that about 30,000 acres are added each year, usually far from cities. Production costs are high because of the expense of pumping water.

Farming consumes about 85% of all water used in the state, but even so agriculture is losing much of its influence over state water policy. In the late 1950s, powerful large landowners in Southern California teamed with industrial and urban interests there to push through the controversial $2.8 billion State Water Project, which essentially diverts water from Northern to Southern California. The project has been heavily criticized for economic and environmental reasons, and environmental groups have forced delay in completing the project, as well as in starting new projects.

State laws have been changed over the years to include greater consideration for the environmental impact of water-diversion projects, says Ronald Robie, vice chairman of the State Water Resources Control Board, which has been increasingly skeptical of such projects. "This is a new age," he says. "The board in the past has tended to be a rubber stamp for agricultural projects."

Environmentalists are pressuring agriculture in other ways, too. The state boasts that its pesticides-regulation law is the toughest in the nation. (But Mr. Chavez of the UFW says provisions to protect workers reentering sprayed fields are "inadequate.") The state also has reasonably tough regulations controlling animal waste disposal.

On agricultural pollution, California's farmers took a strong positive stance. Concerned that the state would ban agricultural burning altogether, farm groups supported a regulated burning law that allows farmers to burn their stubble, vines or whatever only on certain clear days. The farm groups "were quite in tune with the concept that regulated burning would give them a good public image," one state official says.

Farm-group leaders characteristically complain about the "restrictive" nature of environmental laws, but in the long run some observers think agriculture may regain some lost eminence in an atmosphere of growing environmental concern.

"It's true that agriculture has declined as a political power in the state," declares A. Alan Post, the legislative analyst for the state. "But agriculture may come back to power in an entirely different way. There's now more emphasis on environmental concerns, and everybody is thinking about more open space, and agriculture is open space." Adds an aide to Mr. Post: "Environmentalists pretty much accept the premise that use of land for agriculture is a useful and desirable thing."

—WILLIAM WONG

Fading Farm Bloc

The farm bloc is losing its clout in Congress.

Witness a scene at a 1970 hearing of the House Agriculture Committee. Under consideration was the food stamp program, long viewed disapprovingly by the conservatives who control the committee. But when the discussion turned to a proposal by the chairman, Texas Democrat W. R. (Bob) Poage, to require food-stamp recipients to work, John Kramer, executive director of a liberal group called the National Council on Hunger, likened the plan to Soviet-style regimentation.

"Is the chairman for real?" said Mr. Kramer scornfully. "Do you want to transfer the Communist system to the United States as the price for distributing food stamps?"

The open contempt some liberals like Mr. Kramer display toward Mr. Poage and his fellow conservatives on the House Agriculture Committee tells a good deal about the waning power of the once formidable Congressional farm bloc. In the past, city liberals felt obliged to kowtow to conservative rural Congressmen in order to pass welfare legislation like the food stamp program that the Agriculture Committee handles. But now the urban lawmakers think the balance of power has finally shifted; they are increasingly confident that they can force farm-region legislators to bow to their demands.

This showed up in 1970 in an intriguing behind-the-scenes struggle in the House Agriculture Committee, as it pondered in closed sessions what to do about major legislation dealing with food stamps and farm subsidies. The House was the main battleground on both food stamps and farm aid, which the committee had combined in one legislative package. The Senate already had passed a greatly expanded food-stamp program and was expected to accept continued farm price supports.

But the House was antagonistic toward the farm subsidies its Agriculture Committee zealously promoted, while the committee was hostile to the liberalized food stamp legislation that a Congressional majority apparently favored. The panel, after months of indecision over the issues, came under increasingly heavy pressure from the diverse constituencies of both the farm and food stamp programs to quit hemming and hawing and to produce legislation.

Yet the Southern Democrats and conservative Republicans who control the committee were torn between their philosophy and practical political realities. They fervently wanted to expand aid for farmers and limit food stamps for the poor. But their legislative judgment told them they should recommend just the opposite, or else risk having a farm bill crippled or even killed on the House floor while the food stamp program was expanded anyway.

No one is more beset by internal conflict than Chairman Poage. On the one hand, he is the first to say that legislative situations such as this call for delicate diplomacy. "We have got to compromise and make sacrifices and concessions to all points of view if we are going to pass a farm bill," he counseled his committee colleagues.

On the other hand, few have more difficulty putting this advice into practice than Mr. Poage himself. The short-tempered Texan is among the least likely peacemakers in the House. Indeed, during his nearly 40-year tenure in office, he has acquired a certain fame for irritating other House members with speeches he delivers in a raspy roar.

It's not so much what he says as the blaring volume of his voice. "Hell, Bob Poage even aggravates himself with the sound of his voice," says an Agriculture Committee Democrat.

Mr. Poage also found it next to impossible to compromise his life-long concern that "dead-beats and no-good pool hall bums" shouldn't benefit from welfare programs. "I am ready to help all those who need help, but I definitely am not ready to pass out free food or food stamps to those who can but won't work," he declared.

Thus, while Mr. Poage backed a measure that would remove ceilings on food-stamp spending, he insisted on the work-requirement amendment, which many other lawmakers found objectionable. Under the Poage amendment they maintained, children could lose their eligibility for food stamps if their mother or father refused to accept work that a state employment office offered. Some liberals also complained that the Poage work rules would enable Southern states, for example, to force poor Negroes to accept menial jobs as a condition of receiving federal food aid.

The chairman further antagonized some liberals by supporting another amendment, tentatively approved by the committee, that would require states to pay part of the cost of food stamps.

"They're winding up with a bill designed to keep people out of the food-stamp program in the guise of liberalizing it," Mr. Kramer of the National Council on Hunger charged.

Within the committee, Mr. Poage sometimes wavered when outnumbered liberals warned that many House members could be expected to share Mr. Kramer's negative reaction. But other committee conservatives quickly stiffened the chairman's resolve. Once, an insider recalled, Mr. Poage's interest in a liberal member's compromise suggestion ended abruptly after a Republican committeeman upbraided him.

"Bob, you're the only Texan I've ever known to run from a fight," the Republican reportedly said. "My people didn't send me here to vote for social welfare programs and neither did yours."

As urban-suburban strength grows in the House, Mr. Poage and his conservative farm-district colleagues seem less and less representative of the chamber's membership. On the Agriculture Committee are a number of the most conservative men in Congress. The committee's rural Representatives are increasingly fearful of urban-suburban attacks on farm programs, and the chairman often pictures himself as the beleaguered captain of a dwindling band.

Mr. Poage, who spent his childhood on a Texas ranch, noted at the time of the 1970 skirmish that of the 435 House districts, there were only 31 in which at least one-quarter of the population was directly engaged in farming. His own central Texas district is less dependent on cotton growing than it used to be; his hometown of Waco, though still a cotton-marketing center, relies increasingly on industry and military installations.

Chairman Poage and his fellow conservatives are aware that there are many more liberals in the House than on the Agriculture Committee. In 1970, they hoped to pick up needed urban votes by linking the food stamp expansion to the farm legislation. But a number of liberal Congressmen resented this strategy.

The Agriculture Committee is "holding food-stamp legislation hostage so a farm bill can be reported out with hope of passage," contended GOP Rep. Silvio Conte of Massachusetts, a leading opponent of high farm subsidies. "My colleagues will not stand for such a tactic. Indeed, it may backfire and be the surest way to prevent passage of any new farm program."

Passing a farm bill to extend or replace the law that expired at the end of 1970 proved to be an extremely tough task. The House twice voted to place a $20,000 limit on annual payments to any one farmer (some 6,000 farmers received higher payments in 1969), only to have the Senate undo its work. By early 1970, House proponents of strict limits were convinced they had the votes to prevail.

"I predict that Congress will pass no farm bill which permits annual payments to individual farmers in excess of $20,000, even if this means no bill at all," declared Illinois GOP Rep. Paul Findley, co-sponsor with Rep. Conte of the amendment to limit subsidies.

This sort of talk infuriated Mr. Poage and most other Agriculture Committee members. "All these people from the outside who keep telling us how to run agriculture don't realize that a $20,000 limit will wreck the program," the chairman contended. "The big farmers will just increase their acreage if we take their payments away (for not growing crops), and that's bound to wind up hurting the little farmers, probably the Negro farmers in Mississippi that all these people say they want to help."

Still, Mr. Poage realized that "there's a hysteria going on in this country, and I know you can't pass a farm bill without having some kind of limits."

Consequently, the chairman said he was willing to support the sort of subsidy limits the Nixon Administration had proposed. That plan would impose a graduated scale of limits from $20,000 up to $110,000 per crop, with maximums varying according to how much a farmer was then receiving in Government payments. No farmer could receive total payments of more than $330,000. But this would have allowed higher payments than opponents favored.

The Nixon administration didn't push any farm program forcefully, and that was one of the Agriculture Committee's principal problems. During the 1960s, the committee could rely on a Democratic administration to pressure city Democrats to vote for aid to farmers. But with the Republicans in power, many of these city Democrats were balking at voting for any farm bill, and it wasn't clear whether the Republicans could supply enough votes in the House to offset Democratic defections.

Indeed, seven months after the House Committee began mulling farm legislation, there was still deep division among the vitally affected interests. The big American Farm Bureau Federation wanted to phase out farm price supports; a coalition of other farm groups insisted that Government aid should be increased. The administration proposed shifting to a land "set-aside concept" it contended would give farmers more planting freedom; most committee members preferred to stick with the system of annual acreage allotments for major crops— cotton, wheat and feed grains. A key dispute centered on how much discretion the Agriculture Secretary should have in setting price supports.

Mr. Poage, an impatient man by nature, strove to move things along. Though his blustery manner makes him seem a tyrant to some outsiders, committee colleagues say he gives all panel members a chance to shape legislation in the hope of reaching a consensus.

He's in his Capitol Hill office 10 to 12 hours a day, often seven days a week, having curbed an appetite for junkets that used to take him out of the country for long stretches before he became committee chairman in 1966. (Former Rep. Harold Cooley of North Carolina, who used to travel frequently with Mr. Poage, says Lyndon Johnson once gave his fellow Texan a small American flag and told him: "Bob, if you find any place in the world that you haven't already visited, take this flag and claim it as undiscovered territory."

Nowadays Mr. Poage doesn't do much besides tend to Agriculture Committee business and perform chores for his constituents. He is solidly entrenched in his district.

—NORMAN C. MILLER

—Postscript—

After lengthy hassling, the food-stamp and farm price-support issues finally were considered by the House in separate bills. The result in each case was a typical legislative compromise. Agriculture price supports were extended for major crops, but the farm bloc had to accept a $55,000 limit on subsidy payments. Food-stamp outlays were expanded, but Rep. Poage and other conservatives succeeded in enacting a controversial work requirement allowing a cut-off in aid if an adult member of a family refused to accept an available job.—N.C.M.

The Farm Baron

No Cabinet Secretary holds office for two decades spanning five administrations. So it's a little puzzling to the uninitiated when farm-minded politicians and lobbyists talk about the powerful "permanent Secretary of Agriculture" who has held sway since President Truman's time.

But they aren't kidding. They're talking about Rep. Jamie Whitten, a courtly and conservative Mississippi Democrat who has been chairman of the agriculture appropriations subcommittee in the House since 1949. In a very real sense, Mr. Whitten's secure base gives him as much or more influence than any Agriculture Secretary over policies ranging from crop price supports to food aid for the hungry.

Indeed, former Secretary Orville Freeman sometimes complained to associates: "I have two bosses: Lyndon Johnson and Jamie Whitten."

While Messrs. Johnson and Freeman passed from power, the Mississippian expanded his sphere of influence. To the dismay of conservationists and consumer lobbyists, who regard Mr. Whitten as their enemy, his subcommittee acquired budget jurisdiction over the burgeoning array of federal environmental and consumer-protection activities.

"They've really put the fox in the hen house, but there's no way to do anything about it," a GOP Congressman said in 1971.

But some thought his takeover in the environmental field was a desire to do some log-rolling with urban legislators who are big on ecology. They reasoned that Mr. Whitten would be able to win votes for his cherished farm appropriations by promising backing for sizable sums for environmental activities.

There's no doubt, anyway, that he uses his power for his chosen purposes. The Mississippian single-handedly killed an ambitious Agriculture Department rural-development scheme aimed particularly at upgrading the lot of poor Southern blacks. And he has successfully defended old-line farm programs against executive budget-cutters, notably by frustrating the annual efforts of the last four Presidents to slash soil-conservation payments to farmers.

A former department official sums up the clout that Mr. Whitten's strategic position gives him: "There is no way you can beat him when he's dead set against you. What he brings out of committee is what you get."

Mr. Whitten, who also serves on the powerful defense and public works appropriations subcommittees, is one of the most potent of the 13 "barons" who preside over units of the full House Appropriations Committee. Each of these subcommittee chairmen, sometimes collectively known as "the College of Cardinals," wields immense power over the sector of the federal establishment whose funds he controls.

The influence of the mostly conservative appropriations barons largely escapes public scrutiny due to their regal custom of conducting most subcommittee business in secret. Even other members of Congress have difficulty discovering how the barons are shaping agency budgets until money bills are unveiled just a few days before House floor action.

The secrecy is designed to make it hard for dissenters to organize effective opposition on the House floor. And it is especially difficult to beat appropriations barons because of their power to reward or punish other House members seeking funds for pet projects.

It is even difficult to challenge most barons within

the 55-member Appropriations Committee. Except in rare controversies, the full committee functions only to rubber-stamp the decisions of the subcommittees. And with few exceptions, the subcommittee chairmen get their own way within their units.

"The chairmen are all-powerful," said Sidney Yates, an Illinois Democrat who was a maverick member of the committee in 1971. "It's true that the committee can upset them any time it wants to, but it just isn't done."

A major reason is that the barons observe a code that a threat to one is a threat to all, and thus they almost always back one another in a fight. On defense appropriations, for instance, they have banded together to head off cuts urged by House doves.

The barons' alliance is strengthened by the disposition of senior GOP committee conservatives to line up on most issues with the conservative Southern Democrats who head most subcommittees. The Appropriations Committee thus is controlled by a bipartisan conservative group of some two dozen senior legislators headed by Texas Democrat George Mahon, the soft-spoken chairman.

The collective leadership arrangement gives Mr. Whitten and the other barons power surpassing that of most chairmen of the legislative-policy committees in both House and Senate. "A bureaucrat only has to worry about a legislative committee every two or three years when he needs his program renewed, but he has to face the appropriations guys every year," observes a former Johnson administration official. "The annual review gives them great power because the whole bureaucracy becomes attuned to their thinking."

With this power, Mr. Whitten and other barons sometimes kill administration policy initiatives in the bud. A classic example occurred when former Secretary

Freeman quietly began organizing a rural-development program that would have aided Southern blacks. Mr. Whitten was tipped off by department bureaucrats and raised the roof, charging that the department was "misusing funds" and planning to hire 5,000 new employes.

"He stopped us dead," recalls one former Freeman aide. "He managed to reduce the whole program to a token operation."

The incident illustrates the power an appropriations baron has to intimidate the bureaucracy. "The beauty of being the Lord High Executioner is that you seldom have to use the axe," remarks one House member. "All a guy like Jamie Whitten has to do is raise an eyebrow and a tremor runs through the Department of Agriculture."

Career employes know that their future may well be blighted if they cross the normally genial Mississippian. "Jamie is an old prosecuting attorney, and he can really bore in when he wants to show a guy who's boss," says a former Agriculture Department official. "He doesn't do it often, but every now and then he'll hang up a bureaucrat to dry and the message goes out through the system: Don't fool with this guy."

The barons don't always get their way, of course. For example, federal financing of the supersonic airliner was killed despite their strenuous efforts to save it. Liberalized voting rules in the House have made it easier for the rank-and-file to mount successful attacks on the committee's decisions. And more often some members are breaching the committee's long tradition of trying to settle all fights within the panel and having everyone close ranks to support bills on the floor.

Still, the power of the barons remains very strong. Barring a dramatic break with the seniority custom, they can retain lifelong holds on their posts as long as the Democrats control the House. Mr. Whitten has had no Republican foe at all since 1966, and in that year he trounced his opponent by a five-to-one margin. Thus, most barons are about as free from pressure as a politician can be.

Certainly this is true of Jamie Whitten, whose 30-plus years of seniority make him next in line to head the full Appropriations Committee, despite a brand of Mississippi conservatism that puts him at odds with his party's liberal majority. Indeed, Rep. Whitten assumes a stance of total independence.

"I am a representative of the people of (my) district," he says. "I don't believe that Mississippians agree with either the Republican or Democratic Party or with the views of the leadership of either."

Mr. Whitten was elected to the Mississippi legislature when he was only 21 and subsequently served as a district attorney before winning a special congressional election in 1941, when he was 31 years old. When back home, he lives and practices law in the northwestern Mississippi town of Charleston (pop. 2,811), just about 10 miles from his birthplace at Cascilla. In Washington, he shuns the social circuit and puts in 10-hour and 12-hour work days, currently spending his mornings at the defense appropriations subcommittee's meetings and convening his agricultural unit in the afternoons.

What his cotton-grower constituents want from Congress are federal supports for crop prices, development of pesticides to fight boll weevils and other programs to benefit agriculture. For these purposes, they could hardly have a more effective Representative than Mr. Whitten, although he complains that it gets tougher and tougher to persuade urban Congressmen to pay heed to the farmers' wants.

"Since we have gotten the news media and since we have become urbanized, few people realize that life itself is tied to the land," Mr. Whitten says. He considers it his responsibility to overcome that problem. "If somebody doesn't represent the land and the trees and the natural rescurces and see that those who produce the food for the rest of us get a fair shake, there won't be any people to represent," he declares.

A fair shake for farmers, in Mr. Whitten's view, means preservation or expansion of old-line farm programs: price supports, soil and water conservation, rural electrification and rural housing aid. And in his perennial conflict with administration budget-cutters, Mr. Whitten almost invariably wins. He has restored a total of some $2 billion of agriculture conservation payments that Presidents have been trying to cut since the 1950s.

Recognition of the chairman's power has restrained the Agriculture Department's regulation of pesticides, despite growing pressure from environmentalists for a tough policy. "We would have done much more about pesticide regulation if it hadn't been for Whitten's attitude," says a former Agriculture Department official.

Harrison Wellford, an associate of Ralph Nader who has studied Agriculture Department operations, charges that Mr. Whitten has hampered the pesticide-regulation bureau by making sure it was understaffed. Moreover, despite strong protests from conservationists, a federally subsidized effort to kill fireants in Southern states by dumping a controversial pesticide called mirex from airplanes has been continued on Mr. Whitten's insistence, according to Mr. Wellford and others.

Mr. Whitten probably is Congress's most knowledgeable and determined defender of the farmer's right to use pesticides without heavy restrictions. When the late Rachel Carson's book, "Silent Spring," made pesticide use a public issue, Mr. Whitten counterattacked by defending the use of the chemicals in a book of his own, "That We May Live." A number of chemical companies bought many copies for use in their public-relations campaigns against stringent pesticide regulations.

Mr. Whitten says he is just trying to make sure that a "balanced" regulatory policy is followed so that farmers won't have their pest-fighting weapons endangered by "emotional" attacks on the safety of pesticides. "I am convinced that unless the Department of Agriculture stands up to these (antipesticide) drives, the cost of living can greatly increase," Mr. Whitten says. "The health danger would greatly increase, and malaria would be back with us, and many, many other things."

Mr. Whitten's pesticide position is a major reason conservationists were stunned by the decision to give the Mississippian appropriations jurisdiction over the new Environmental Protection Agency. But some worried environmentalists see evidence for the theory that his major goal is log-rolling to preserve farm programs.

As backing for this theory, some people cite his support of food-stamp appropriations in 1971 despite initial hostility. When hunger became a national issue in the late 1960s, Mr. Whitten infuriated liberals by assigning FBI agents on loan to his committee to investigate people who had told TV interviewers about hunger in Mississippi and other states. His counterattack on the hunger issue was so strong that he was denounced as a man who "has anesthetized his soul to human misery and indignity" in the book "Let Them Eat Promises" by journalist Nick Kotz.

But Mr. Whitten hasn't tried to cut food-stamp appropriations as spending soared to well over $1 billion a year under the Nixon administration. Other politicians attribute this flexibility to a Whitten decision that fighting food stamps would only invite liberal attacks on farm appropriations.

Many lawmakers who disagree sharply with Mr. Whitten on several issues nonetheless defend him against critics such as Mr. Kotz. "The criticism of Jamie Whitten by some liberals has gone far beyond the realm of propriety and accuracy," says Colorado Democrat Frank Evans, a liberal member of the Whitten subcommittee. "Jamie Whitten always deals with you fairly," adds GOP Rep. Paul Findley of Illinois, who has frequently clashed with the Mississippian on crop-subsidy issues.

And even Mr. Whitten's most bitter critics agree that he brings to his powerful post a sharp mind and a willingness to work long hours mastering tedious budget details. "That's a big reason he's so hard to beat; he really knows his stuff," says a liberal lobbyist.

—NORMAN C. MILLER

POSTSCRIPT

Rep. Whitten has fulfilled the expectations of those who looked for him to use his control of environmental appropriations to do some log rolling. On the one hand, he has permitted steadily rising outlays for environmental programs. On the other, he has used his power to restrict some environmental-enforcement activities even while successfully bargaining with ecology-minded legislators for their votes backing his cherished farm programs.—N.C.M.

Secret Study

Agriculture Secretary Earl Butz's spirited defense of the way the Nixon administration handled the Russian wheat deal was jolted by the disclosure that his department suppressed a report that might have helped U.S. farmers get higher prices for their wheat.

The report, an analysis of Soviet crop conditions completed in mid-August 1972 by the Agriculture Department's economic research service, concluded that the outlook for Russian grain production had worsened from an earlier estimate. But the study was never released because the research service deemed the findings "too controversial" and classified the analysis "confidential."

Publication of the study, of course, might have provided another dose of bullish news for already surging wheat markets. Farmers with wheat to sell presumably would have benefited from this. But exporters faced with the need to cover short-term foreign deliveries with cash-market purchases wouldn't have been happy. The Agriculture Department itself, debating at the time whether to continue increasing wheat-export subsidies to exporters in pace with rising domestic market prices, would have found its own problems greatly compounded.

The disclosure of the report casts doubt on Secretary Butz's assertions that the Agriculture Department did everything possible to alert U.S. producers to the magnitude of Soviet grain needs. Statistics are lacking, but a sizable number of growers at the southern end of the winter-wheat belt (where the harvest starts) apparently sold their 1972 crop too soon to benefit from the

30% jump in prices that resulted as the full extent of Russian buying surfaced.

"Some of the guys in Texas and Oklahoma are quite upset over missing the price," Jerry Rees, executive vice president of the National Association of Wheat Growers, said. Adding to the disappointment was the provision in farm law that shrunk the government subsidy to producers as market prices rose—a feature estimated by the National Farmers Union to have lopped $100 million from payments in the 1972-73 marketing year. Thus, a farmer who sold his wheat before the price runup also got a smaller government check than he anticipated at planting time.

Earlier charges of inadequate information from the Agriculture Department were made by John Schnittker, who was Under Secretary of Agriculture during the Johnson administration and who then became a Washington commodity consultant. Mr. Schnittker claimed that the U.S. agriculture attache in Moscow filed reports in June and August that disclosed more about the Soviet grain situation than was known in the U.S. at the time—but the department didn't release them.

Officials in the department's foreign agricultural service confirm that such attache reports indeed are being held in a "confidential" file, separate from the bulk of attache reports available to the public. The confidential reports, in fact, along with weather information from satellite observations and other data, were used by the department's economic research service to prepare the "special highlight" analysis dated Aug. 18. It was this study that projected a further decline in Russian production of wheat and other food grains from estimates contained in a similar analysis done in July.

But while the research service had released the July study to the public, it elected not to do the same for the August one, circulating it instead only among a select group of top department officials. Those with knowledge of the August study say the decision against making it public was partly influenced by the belief of still other analysts in the foreign agriculture service that the new estimate was too tentative to be reliable. Yet the biggest factor appears to have been fear that the August analysis was, as one economic research service man put it, "too controversial" in view of turbulent wheat-market conditions at the time.

To understand this reasoning, it's useful to reconstruct the wheat-market situation in the weeks before and after the Aug. 18 analysis. The chronology, to the extent it's known:

—June 28. The newly arrived Soviet delegation to negotiate a grain deal notifies U.S. officials of its willingness to accept Washington's basic terms for Commodity Credit Corp. financing of Russian grain purchases. At the time, the domestic price of No. 2 hard red winter wheat at Gulf ports stands at $1.64 a bushel, with the department offering to pay a one-cent subsidy to exporters. That's in accord with its established policy of maintaining U.S. wheat at approximately $1.63 in the world market for competitive reasons.

—July 8. The $750 million, three-year deal between the two countries is announced. Secretary Butz predicts U.S. sales actually could exceed $1 billion (the figure for the first year alone was more than that), but he and other officials indicate that it's mostly corn and other feed grains that the Soviets want, rather than wheat.

—June 28-July 21. A Soviet grain-buying mission, which arrived with the agreement negotiators, quietly places orders for roughly 250 million bushels of wheat to be shipped before next June 30. By July 21, the Gulf price is up to about $1.76, and the subsidy is 13 cents.

—July 29 to about Aug. 20. After having returned

home, the Soviet grain buyers suddenly reappear in New York and resume negotiations with grain companies for still more wheat.

—Aug. 1. With the Gulf price up to about $1.80, Agriculture Department officials grow concerned about rising subsidy costs and the fact that the expanding subsidy may be pushing, rather than following, the domestic price. They decide to let export wheat move above the $1.63 line by allowing the subsidy to lag behind domestic prices.

—Aug. 8. Southwestern Miller, a trade publication, prints a report that Soviet wheat purchases will reach 400 million bushels in the 1972-73 marketing year, an amount almost equal to two-thirds of total wheat exports in 1971-72. Department officials, claiming they didn't have the news before anyone else, later say they're pleased but mystified by the unexpected orders. (A few department experts wonder privately, though, if the July 8 grain agreement was a smoke screen; the majority of Russian buying has been in wheat, not feed grains, and mostly for cash, not credit.)

—Aug. 24. With the Gulf price up to about $2.14 and the subsidy a fat 38 cents, the department advises exporters that it's going to permanently abandon its "commitment" to maintain the $1.63 world price. The decision reflects Nixon administration concern that the subsidies' effect on domestic prices could lead to embarrassing bread-price boosts.

—Aug. 25. At a meeting with grain-trade representatives to which the press is reluctantly admitted, Assistant Agriculture Secretary Carroll Brunthaver announces the policy change. Apparently influenced by exporters' protests, though, the department offers an unusual nine-cent retroactive supplement to the 38-cent subsidy to cover export sales made through Aug. 24 and registered by Sept. 1. Some 280 million bushels qualify for this provision, representing almost a 60% jump in the wheat registered since July 1 and lifting estimated subsidy costs to almost $250 million in the first two months of the 1972-73 marketing year. (Wheat subsidies in 1971-72 amounted to only $65.4 million, and the high was $160 million in the 1965-66 crop year.)

When the economic research service put out its first special highlight appraisal of current Soviet grain crop prospects, on July 14, it found that Russian food-grain production, primarily wheat, would be down "roughly" 20 million metric tons from the 113 million tons harvested in 1971. (A metric ton is approximately 2,200 pounds.)

The July report didn't have any discernible impact on grain traders, however, apparently because it merely confirmed Soviet supply needs already evident from the actions of Russian buying missions in Canada, the U.S. and elsewhere. The general expectation at that time was that Russia would have to import 14 million to 15 million tons of grain for itself and its East European allies, including the approximately 250 million bushels, (or seven million tons) negotiated in the first round of purchases in the U.S.

Reexamining the situation a month later, the economic research service decided to lower the Soviet food-grain production figure by another seven million tons, to 86 million tons. The reduction apparently reflected the attache's opinion that severe cold, drought and finally record-breaking heat in the Ukraine and other Soviet winter-wheat regions had caused more damage than initially believed.

Critics of the administration contended that big grain exporters knew by the time of the study's completion that the Soviets actually had committed themselves to buying some 11 million tons of U.S. wheat, rather than the seven million originally calculated. But most farmers probably didn't have any idea the Russian commitment was that big, these critics say. They say issuance of the analysis would have helped to spread the news more quickly.

—BURT SCHORR

Befuddled

Telling about a farmer who because of the farm boom in 1973 was able to pay off all his bills and still have money left over, Earl Butz, the Secretary of Agriculture, comes on like a country preacher.

"You think he isn't happy? And his banker? And the school board and the church he belongs to?" Mr. Butz asks rhetorically. "The man is proud of being a farmer. He's got his head held up high—and he walks down the street with an air of confidence, and money jingling in his pocket. And it's about time!"

Others in the Nixon administration, however, didn't fully share that glee. Talks with presidential advisers, Treasury, Commerce and State department officials, Congressmen, a bevy of outside experts (and many Agriculture Department aides) showed they were worried and a little confused about the farm boom and how to handle it.

For, while Mr. Butz was gloating about money jingling in farmers' pockets, others in the administration had to try to explain to food shoppers why there was less jingling in theirs. Then, to some, there was the haunting possibility of the boom turning into a bust—with new farm surpluses, new burdens on the taxpayer and with food prices, while temporarily easier on shoppers' pocketbooks, low enough to discourage farm production.

Such pessimism wasn't unwarranted, because in 1973 farmers were barely able to produce enough to match domestic and foreign needs, and food prices surged.

The Nixon administration opted for full steam

ahead on production and exports in 1973-74. Idle acres were returned to production, and farmers were given incentives to grow all the crops they could, while Agriculture Department attaches around the world were given the word to go all-out in boosting sales. As simple as this policy sounds, it had potential pitfalls that were readily conceded by some policymakers.

Carroll G. Brunthaver, who was in 1973 the assistant agriculture secretary in charge of international affairs, said matter-of-factly: "This is fun, unless we're wrong." If that happens, the cost to the taxpayers could be "fantastic," he adds.

The policy was—and still is—a gamble because if high production and high sales don't go hand-in-hand, the whole thing would backfire. If farmers produced robust harvests and overseas sales fizzled, the government would face a mind-boggling subsidy bill. (Domestic food prices, however, could be cheaper.) If farmers should have an off-year and foreign demand zoomed, U.S. consumers could face food shortages and higher prices. And government officials would have to agonize about imposing export controls.

"This isn't a well-thought-out policy," a staffer for the Senate Agriculture Committee said. "It's a defensive action to handle today's problems. The administration hasn't come close to formulating a long-term agriculture program."

Enough contradictions and maladjustments abound to confirm that assessment, some administration officials privately conceded. Some examples:

—Despite outcries ,for higher production, the administration whittled away the Agriculture Department's research budget. In 1973, the amount allocated at the federal level for finding new ways to produce more crops totaled $183 million, down $17 million from the previous year.

—The department's manpower was cut by about 3,-000 persons in 1973 to 79,000 full-time employes, but critics say it's still topheavy with experts on running outdated crop-subsidy programs that have cost taxpayers $184 billion since 1933. With farm profits at records, Iowa corn growers and Kansas wheat harvesters didn't need much help from Washington, critics argued.

—According to critics, government has been slow to consider setting up buffer stocks to protect against major market fluctuations overseas, such as Russia's sudden sharp demand for American grains. Sen. George McGovern has called for a crop-stockpiling program. "We have missiles and bombs stockpiled for defense," says the South Dakota Democrat. "I don't see why we don't do the same with food." Many government officials agree, and President Nixon indicated to subordinates that he might accept some food-reserve plan. But no action was forthcoming from a Congress that was increasingly preoccupied with the Watergate scandal.

—From the consumer's viewpoint, the administration was too slow to lower barriers to imports of short-supply dairy products. The government had allowed a record amount of foreign powdered milk into the country in 1973, but the Cost of Living Council said this had had negligible impact on holding down domestic prices.

One explanation for this lack of cohesion in farm and food matters was the new attitude in Washington, expressed best by Gary L. Seevers, a member of the President's Council of Economic Advisers: "Agriculture has become far too important to be left to the agriculturists." It also explained how a variety of top-level officials who hardly knew a sow from a silo suddenly found themselves taking decisive roles in revamping agriculture policies to deal with the boom.

Consider the case of George Shultz, who was then Secretary of the Treasury. Usually preoccupied with dollar devaluations, price controls and such matters, he grumbled that never before had he known what a "piggy sow" was. But he quickly learned that a piggy sow is a pregnant pig and that hundreds of them were being slaughtered in mid-1973 because farmers feared skyrocketing feed-grain costs and meat-price controls would make it economically impossible to raise the piglets or continue feeding the sows.

To Mr. Shultz, this development portended higher pork prices, consumer anger, foreign fears of U.S. export controls and perhaps even a how-come call from the White House.

Mr. Shultz's new awareness indicated how the new farm situation churned up a flurry of role-changing and reassessments. But there was by no means final answers on what the turnabout from crop surplus to scarcity meant for the longer term and what the government could do about it. "Let's just say we're, well, a little befuddled," a Treasury aide said in the autumn of 1973.

Indeed, ask most policymakers about agriculture and they'll lace their language with mention of the U.S. balance of trade, the energy crisis, population control, changes in dietary habits and other subjects that on the surface appear to have little to do with planting corn in Iowa. Few officials outside the Agriculture Department talk much about the farmer himself.

And inside the Agriculture Department, "we're hearing from a different group of people," says Don Paarlberg, an assistant agriculture secretary. "We don't hear from the farmers any more, we're hearing from the consumers."

And that raised problems because the Agriculture Department is geared for the welfare of the farmer and not for that of the consumer.

When a consumer group marched on the Agriculture Department demanding that Secretary Butz resign, an administration official confided, "We got off easy. They should have been marching in front of the White House or Treasury; there's where the decisions are being made."

In fact, most major policy pronouncements affecting the farmer in 1973 had come from the White House or the Treasury. These included the lid on meat prices (openly opposed by Mr. Butz), the temporary cutback on exports of soybeans and related products (again opposed by Mr. Butz) and the setting-up of the Council on Economic Policy to study the food situation (chaired by Mr. Shultz). The milestone farm law enacted in 1973, which set up crop-price incentives for full-blast production, was drawn up in Congress and at first was opposed by Mr. Butz as inflationary.

"Earl has been reduced to the role of a cheerleader for the farmer," an administration aide observed. "Oh, don't get me wrong, he's doing a damn good job; he talks their language." One speech, in which Mr. Butz predicted record farm income for 1973, was entitled "Out of the Wilderness Into the Promised Land."

If farmers are to continue making what Mr. Butz called "a decent income," Americans have to get used to earmarking more of their take-home pay for food, the Secretary and other champions of the farmer said. "Americans have been getting a bargain for years," Sen. McGovern asserted. "Well, the bargain days are over."

Mr. Brunthaver of the Agriculture Department pointed to what he considered the bright side, the farm-subsidy drop that is a boon for taxpayers. "For the first time in years, the farmers are getting their money from the marketplace, not the Treasury," he said.

Yes, critics agreed, but primarily because farm-product exports, which soared 60% in 1973 from 1972, put the squeeze on domestic supplies.

Despite that, the Nixon administration shied away from further farm export controls. State Department diplomats warned that any more interruptions in exports, like 1973's temporary clampdown on soybean shipments, would send foreign customers scurrying to find more suppliers. Export controls also would upset efforts by the administration to negotiate a reduction in foreign trade barriers to U.S. agricultural products. Agriculture Department men added that the future tempo of shipments abroad will largely determine whether the farm boom will turn into a bust.

"This is why we have to go easy when making any major changes in our agriculture program," Mr. Paarlberg insisted.

—MITCHELL C. LYNCH

Part Five

FARMING AND THE ENVIRONMENT

The environmental movement had gotten rolling at quite a good speed before it noticed the pollution coming from farms. It was as though it hadn't occurred to anybody that agriculture had its ecological drawbacks. Farmers themselves were surprised (before many of them became angry, like the captains of industry before them)—after all, wasn't modern farming the ultimate in natural-resource husbandry? Well, not quite, as it turned out. As with industry, the costs of great boosts in productivity were sometimes measured environmentally—deaths from pesticides, runoffs of chemical fertilizers and manure into streams, and so on. But for the most part, farmers and researchers have met the criticism with efforts to minimize and clean up agricultural pollution.

Rural Contamination

Cities are filthy. Farms are clean.

That's the way it seems to most people. While the battle over insecticides may have marred the farm's wholesome, bucolic image, the American farmer has managed to stay aloof from much of the national furor over pollution.

But all this is changing. Pollution controllers increasingly conclude that farms account for a surprising amount of the nation's pollution in several categories —even in cities. For instance, many cities draw their drinking water from rural areas, where contamination from livestock feedlots and fertilizer-soaked fields is heavy. The National Wildlife Federation, a private conservation group, estimates that fully 15% of all U.S. water pollution emanates from agricultural sources.

Thus, farm critics who once focused mainly on the use of insecticides now question such matters as the way farmers fertilize their crops and operate their animal-feeding facilities.

"There's no doubt that we have a serious problem here, and that answers must be found soon," says Cecil Wadleigh, a chemist who is an adviser to the U.S. Department of Agriculture.

Experts think they are on the brink of solving some of these problems. However, many solutions being proposed could dramatically alter the way farmers raise crops. Indeed, some of the farm methods coming in for the most criticism are the very ones that have enabled U.S. farmers to boost their productivity seven-fold in the last 50 years. As a result, solving the pollution problem could boost the price consumers pay for their food.

The damage caused by farm pollution is clear. For instance, the nitrogen compounds entering lakes and rivers from feedlots and chemically fertilized fields have fouled recreational facilities and water sources used by farmers and city dwellers alike.

The nitrate count in the Decatur, Ill., city water supply reached 60 parts per million in April 1971, well above the safety limit of 45 parts per million set by the U.S. Public Health Service. Health officials in the area say the count is typically highest in the spring, just after most farmers in the area fertilize their fields. In addition, the Sangamon River, from which Decatur and surrounding areas draw their water, is a fertile breeding ground for algae, which at certain times of the year give the drinking water a foul taste. The algae thrive on chemicals that drain into the water from the fertilized fields.

The fertilizer can also make water plants multiply so fast that they choke rivers and streams. When the plants decay, they consume large amounts of oxygen, which often makes the water uninhabitable for fish and other aquatic life.

Officials are moving to control some forms of water pollution that originate from farming. One is the waste-laden run-off from livestock feedlots that contaminates sources of drinking water. In such feedlots, which are sometimes called, "fattening-up stations," food is brought to the animals. As the livestock don't roam pastures, feedlots concentrate a large number of animals in a small space.

Feedlot operators typically dump the animal waste they collect into ponds, where it decomposes for later use as fertilizer. The problem arises when heavy rains cause the ponds to overflow and run into nearby streams and lakes. This problem has grown more serious as the number of large feedlots has grown in recent years. In Kansas a few years ago, feed-lot pollution was blamed for massive killing of fish in the Arkansas and Cottonwood rivers.

In the last nine years, nearly every major agricultural state has moved to tighten its regulations on feedlots and to ensure that feedlot ponds can handle heavy rains. In some cases, these regulations have proven expensive for farmers. One Michigan feedlot operator says he lost about $100,000 in revenues while his facility was being altered. A few lots located on terrain that precluded expanding feedlot ponds have been forced to close.

But the laws are getting results. "By 1973, all the cattle on feedlots in the state of Kansas will be in approved facilities, and the problem will be licked," Melville Gray, chief engineer for the state's health department, said late in 1972. "That's really something when you're dealing with a pollution problem," he added.

But some other farm problems are still real sticklers, officials concede. One is the runoff of nitrogen-based fertilizers into the nation's waterways, such as occurs near Decatur.

The problem is that no other substance matches the crop-nourishing properties of nitrogen. Experts estimate that without this prime fertilizer, at least 20% additional acres of land would have to be pressed into production to keep up the pace of U.S. food output.

Nobody seriously advocates barring fertilizers from farms. But many scientists contend that farmers can reduce the nitrogen runoff from their fields through more careful use of fertilizer. "If farmers can learn to use fertilizer correctly, much of the problem can be eliminated," says one U.S. Department of Agriculture official.

—FRANCIS L. PARTSCH

Death in the Fields

Juan Hernandez felt fine one morning despite the chemical spray drifting into the orchard where he and fellow laborers were picking lemons. Farm workers around Woodville, Calif., are accustomed to pesticide sprays that blow on them from an airplane or a tractor rig.

By afternoon, however, the stomach pains began, and Mr. Hernandez (that's not his real name) vomited in the field. He went home at 6 p.m.; soon his arms and legs grew numb. By 9 p.m., when he was rushed to the Salud Medical Clinic, a one-doctor institution in Woodville, he was struggling to breathe.

A massive dose of atropine—the antidote for nerve-gas poisoning—probably saved his life. The spray that had entered Mr. Hernandez' nervous system through his skin and lungs was an organic-phosphate pesticide. This group of chemicals emerged from World War II nerve-gas research and now comprises the leading U.S. pesticides.

Mr. Hernandez, who disobeyed doctor's orders and returned to work in two days because he needed the money, was one of the fortunate victims. In North Carolina, a 16-year-old boy collapsed in a field where he was picking tobacco. He was left unattended for three hours at the farm manager's home, suffered a brief heart stoppage and died three days later. The medical report listed the probable cause of death as parathion poisoning. Parathion, one of the most lethal and widely used organic-phosphate pesticides, had been sprayed on the field 12 days earlier.

As these cases suggest, organic-phosphate pesticides are causing major controversy. The highly toxic bug killers now account for half of all insecticides sold, becoming even more popular as a result of the federal ban on DDT.

Their effects on bugs are impressive, although not to the bugs. "They're the atomic bomb for insects," one major corporate farmer in California declared enthusiastically. Precise statistics on usage of organic phosphates and resulting illness are difficult to obtain. But in California, which uses 20% of the nation's pesticide output, there were 216 "systemic poisonings" from agricultural chemicals in 1968. Four out of five cases involved organic phosphates. Public-health reports in California list 15 fatalities from organic phosphates among persons in farm work and related occupations between 1951 and 1968. One federal official has estimates of 200 pesticide deaths a year in the U.S., most attributable to organic phosphates.

Until recently the lethal properties of organic phosphates, especially to crop-dusting pilots and others who handle the raw chemical, obscured what some health experts see as a more disturbing and widespread problem. This is the possibility that great numbers of ordinary farmhands may be doomed to chronic, low-level poisoning. Especially affected may be migrant farm workers who pick crops such as oranges and peaches. Some studies confirm suspicions that official reports vastly underestimate illness among farm workers caused by organic phosphates.

One study by the California Health Department in Tulare County, for example, indicated that as many as 150 out of every 1,000 farm workers suffer symptoms indicative of pesticide poisoning. The results of the study are disputed but undoubtedly thousands of farm workers are regularly poisoned without appearing in official statistics. Many periodically suffer what people around

Woodville term "orange pickers' flu"—a collection of symptoms such as nausea and headaches, thought to be caused by organic-phosphate pesticides.

The growing use of organic phosphates thus may become one of the sad ironies of the ecological movement. As DDT and related chlorinated-hydrocarbon pesticides were banned from one U.S. crop and then another, environmentalists praised each restriction as a major victory. DDT and its relatives are "persistent" pesticides, which remain and concentrate in the food chain, eventually poisoning wildlife and perhaps posing a danger to humans who eat foods with high DDT concentrations.

Pesticide makers and farmers were happy to have the organic phosphates on hand as substitutes for DDT. The organic phosphates break down quickly after being sprayed, apparently disappearing long before food reaches the consumer.

But as chemists knew well and farmers gradually learned, the organic phosphates are lethal before the breakdown process. Several drops of some types on the skin can be fatal. Leaves sprayed with many organic phosphates will spread the poison to anyone handling them even weeks after application. DDT, by contrast, is relatively safe for farm workers to handle.

The replacement of DDT by the organic phosphates appears to many farm workers as a poor trade. "While this (the DDT phase-out) may be a blessing for the birds and fish, it's at the expense of the farm worker," asserts Charles Farnsworth, a lawyer and pesticide expert for Cesar Chavez and his United Farm Workers Organizing Committee.

Whatever the arguments for DDT, organic phosphates appear to have the upper hand. The phosphates have been used since the late 1940s, but sales have spurted in recent years as pests became resistant to

DDT and critics challenged the long-term effects of DDT on the environment.

On June 15, 1971, California became the first state to clamp heavy restrictions on farm operators who use organic-phosphate pesticides. It set mandatory intervals between spraying and the reentry of farm workers into a field. The intervals range to 45 days after heavy spraying of the most common and troublesome pesticide, parathion.

Pesticide makers and farm operators are cooperating with the tough California regulation and with a nationwide program coordinated by the U.S. Agriculture Department to inform users of the hazards of organic phosphates. In some cases the motive is to avoid even more stringent regulation—perhaps an outright ban— if the controversy continues. "We're in favor of reasonable safety regulations, but California's rules don't qualify," said one major corporate farmer that was notably silent at a hearing on the regulation.

One thing farm scientists generally agree upon: There isn't any economical substitute for organic phosphates. If they were banned, "the country could support about as many people as it did when the Indians inhabited the continent," claims Gordon Snow, a special assistant in the California Agriculture Department.

Many farmers are seeking to reduce their usage of the more lethal types of organic phosphates and are testing other methods of pest control, such as "biological control"—the use of predators, such as ladybugs, to destroy harmful insects. "I'm sure we've reduced the amount of organic phosphates we use over the past several years," says William Balch, executive vice president of Hegglade-Marguleas-Tenneco Inc., a Bakersfield, Calif., subsidiary of Tenneco Inc., which farms 40,000 acres in California and Arizona.

Much of the pressure for stiffer regulations was

generated by Mr. Chavez and his Farm Workers Union. But some union officials think the law isn't enforced and has no teeth.

The willingness of farm workers to police any system by reporting accidental contact with organic phosphates or related illness is open to question, some observers say. "Farm workers often don't report poisoning because they may lose their job," says Dr. Erwin Brauner, an internist and head of a Tulare County medical society committee that has studied pesticides.

A visit to California's lush agricultural heartland in the Central Valley makes clear the problems that health officials and legislators are up against. One Mexican American worker, a short, friendly man surrounded by a half-dozen children in front of his unpainted house, reluctantly tells a reporter that he has been working in orange groves and was sprayed that day for the third day in a row. "I have headaches all the time," he relates, and he expresses grave doubts that the state will ever be able to enforce the ban on worker reentry. "It won't work as long as it pays to have a crop picked right away," he says.

The high incidence of careless or inadvertent spraying is documented in a study financed by the Field Foundation of New York and conducted by Mrs. Wendy Brooks, a sociologist and wife of the doctor who ran the Salud Clinic in Woodville until it closed. She followed 57 families for a period of a year, providing free medical care at the clinic for them in exchange for monthly blood samples and information on accidental sprayings or other signs of pesticide exposure. She found that nearly a third of the 301 persons who completed the test were accidentally sprayed in the fields during the year. Symptoms of pesticide poisoning occurred in 30% of the adults sprayed.

Some pesticide makers and farmers believe the hazards of organic phosphates are being exaggerated. "Personally, I think the safety record is darn good on pesticides," says J. Marshall Magner, chief entomologist in market development for Monsanto Co., St. Louis, a leading parathion producer. "I wouldn't want to say illness reports (attributed to organic phosphates) are phony, but I suspect some of the reported illness is psychological." He adds quickly, "I'm not trying to play down the hazard."

Scientists at Del Monte Corp., San Francisco, a major food grower and packer, agree that publicity about pesticides could lead to inflated poisoning reports. "I suspect there was a lot of pesticide poisoning unrecognized before," says Duncan Carter, entomologist at Del Monte's research center in San Leandro, Calif. "But now a farm worker with a headache immediately gets a lot of attention."

Mrs. Brooks responds angrily to such comments. The number of poisonings, she says, is grossly understated "because of a lack of knowledge and motivation among doctors and a lack of laboratory facilities."

Del Monte scientists and others confirm the relative lack of medical attention to organic-phosphate poisoning. "We found many rural doctors who lacked information about how to treat it," says Mr. Carter at Del Monte, which distributes a manual to physicians on diagnosis and treatment.

—HERBERT G. LAWSON

Bugs vs. Bugs

Thousands of tobacco budworms, each about two inches long, make their way through the lush foliage of the cotton field, feeding on cotton bolls. Every few days, a torrent of pesticide saturates the field, wiping out entire populations of some insect species, including the budworm's predators. But unlike its enemies, the budworm is strangely impervious to the poison.

In a field free of predators, the budworm flourishes, feeding until nature signals it to reproduce. Soon the field teems with a new generation of the pest, 10 times as plentiful, even more resistant to pesticides, and just as hungry as the last generation.

The budworm will repeat this cycle at least two more times before the cotton-growing season is over. In just a few months, it will have devastated the entire cotton crop.

To cotton growers in the lower Rio Grande Valley of Texas, such an invasion by the insecticide-resistant budworm is frighteningly familiar. Since 1968, this cotton and tobacco pest has repeatedly fought off some of the heaviest and costliest chemical attacks farmers have ever mounted against an insect. The worm has damaged hundreds of thousands of acres of cotton in Texas and destroyed the cotton industry in northeastern Mexico.

When a pest like the budworm is immune to all known poisons, "about the only way to kill him is to run over him with your truck," says Jerry Young, who has grown cotton in the Valley for the past 24 years.

But now the budworm may have met its match. For the past few seasons, growers and agricultural scientists in the Valley have been wielding a new weapon against the insect. It promises to control the budworm and other pests far more cheaply than the drenching pesticide attacks. To the delight of environmentalists, it involves the use of only a fraction of the poisonous chemicals ordinarily used on cotton.

Known as "integrated control," the new approach involves carefully scheduled and limited applications of insecticides. Sprayings are timed with high precision in order to leave friendly bugs alive and free to combat damaging pests like the budworm.

The effort has implications beyond the Rio Grande Valley and beyond the cotton industry. After more than 20 euphoric years of almost insect-free farming, many entomologists believe the heyday of chemical insecticides has passed.

The budworm, they say, is only one of 230 insect species that have already developed a resistance to at least one pesticide—and it's only a matter of time before others follow. Because of their abundance, diversity, and adaptability, insects have always held their own against their brainier competitor, man. Through natural selection, an insect can eventually evolve strains resistant to each of the chemicals man devises to kill it—sometimes in the space of only a few years.

To a large extent, the problem of insect resistance has been obscured by public concern over the hazards of pesticides persisting in the environment. Yet, according to some entomologists, the growing ineffectiveness of chemical weapons in controlling insects may prove even more disastrous for a world trying to feed and clothe an exploding population.

"The continued development of insecticide-resistant strains of insects threatens to render obsolete much of our present pest-control technology," warns Perry L. Adkisson, chairman of the entomology department at Texas A&M University at College Station, Texas.

Farmers unwittingly played into their enemy's hand when, in desperation, they poured excessive quantities of chemicals onto their fields to attack invading swarms of pests. At one time, insect control in the Valley could be managed with less than five pounds of chemicals per acre. Then it jumped to more than 20 pounds an acre as the budworm became more resistant. Such insecticidal "overkill" disturbed the delicate ecological balance between insects and their predators and allowed pesticide-resistant strains to multiply.

"It's a battle as frustrating as the Vietnam war," says Mr. Adkisson, who has studied the Valley's problems for a number of years. "The insects are using guerrilla tactics, and we're using bomb-busters on them— and it simply isn't working."

During the peak years of pesticide production, farmers simply switched to a new poison when resistance problems popped up. But now, the high cost and time involved in developing a new product have caused many chemical concerns to discontinue or cut back pesticide operations in search of a faster return on their research dollar. It's estimated that it now takes about seven years and $5 million to $10 million to put a new pesticide on the market. The result: Few new products are finding their way onto insect-ridden fields.

At the same time, other advances in agricultural science have helped create a brand new set of pest-control problems. By and large, plant varieties have been bred for high yields, with pest resistance characteristics minimized or ignored. Agricultural monoculture, the planting of vast acreages with a single variety of crop, has removed the ecological diversity necessary to hold nature in balance. And advanced methods of fertilization and irrigation have produced larger, healthier plants, which are more susceptible to rapid buildups of insect populations.

The Rio Grande Valley's troubles began in the late 1950s, when the destructive boll weevil began to develop resistance to the DDT-like pesticides that long had kept it under control. To attack the weevil, farmers switched to the organophosphorus pesticide, methyl parathion. Since parathion isn't persistent, the farmers had to apply it more often. Moreover, since it's a "broad-spectrum" insecticide, it isn't selective, and it killed pest predators and parasites.

As a result, by 1962, cotton growers were confronted with an outbreak of two "secondary pests," the bollworm and the tobacco budworm. Unchecked by natural enemies, these pests proved even more damaging to cotton crops than the weevil.

Disaster struck in 1968, when the tobacco budworm showed the first signs of resistance to parathion. Many cotton growers treated their fields 15 to 20 times and still suffered severe losses in yield. (In 1967, Valley farms produced an average of 601 pounds of cotton an acre; in 1968 yields dropped to 420 pounds an acre.) Some growers were able to salvage their crops, but the cost of repeated insecticidal applications—which sometimes ran as high as $70 an acre—wiped out their profits. By 1970, the budworm had grown so resistant that the Valley suffered the lowest yields recorded there in 25 years, with barely 300 pounds of cotton produced per acre.

Two hundred miles south of the border in northeastern Mexico, severe outbreaks of resistant budworms destroyed the entire cotton industry in just three seasons. Planted acreage in the Tampico-Mante region shrunk from some 500,000 acres in 1966 to only 1,000 acres by 1970.

During the boom years, most of the world's major

cotton companies were represented in the area, and 42
cotton gins operated there. Now the gins are boarded
up, and 250,000 laborers are without work. Growers financed by cotton companies lost their land through
foreclosure; fields that once produced $100 million of
cotton each year now lie barren and idle.

"Modern cotton production has been based on an
insecticidal crutch," says Perry Adkisson. "When that
crutch breaks, the industry is in trouble."

As an alternative to that crutch, entomologists
working in the Valley are introducing integrated control. This approach, which is gaining wider favor in
areas where chemical control has failed, calls for the
combined use of insecticides, beneficial insects, and in-
sect-resistant crops. In essence, it restores the balance
of nature to areas of ecological anarchy.

Using integrated control, farmers in the Valley
have kept cotton pests at bay, while cutting their use of
insecticides in half. Conly and Gus Bell, for example,
who grow 1,200 acres of cotton, averaged only seven applications of chemicals in 1971. Before, they had
sprayed as often as 16 or 17 times. "Our costs were down
about a third," says Conly. "In one field (in 1971), we
didn't use any insecticides—and it was the best cotton
we had." Even with fewer sprayings, average yields in
the Valley reached 530 pounds an acre in 1971—among
the highest in years.

Entomologists at Texas A&M estimate that insecticide use on cotton in most areas of the U.S. could be cut
in half without reducing yields. Their projections are especially significant, since nearly half the 138 million
pounds of insecticides applied to U.S. crops each year
are dumped on cotton. What's more, integrated-control
methods in other regions have helped reduce insecti-
dal use on a number of other crops, including peanuts,
grain sorghum and certain fruits.

During the early part of the growing season, the control program carried out by the Bell brothers and other cotton farmers sounds deceptively simple: They abstain from spraying as long as possible, to keep "friendlies" alive to attack the budworm and other pests.

Unfortunately, the ecosystem of the cottonfield contains several other insect pests that complicate natural control of the budworm. If, for example, the boll weevil or fleahopper invades their fields early in the season, the Bell brothers have a painful choice: They can either refrain from spraying and suffer crop losses from the weevil and fleahopper; or they can attack the pests with insecticides, knowing at the same time they are killing the "friendlies" that keep the budworm under control. Under these conditions, the Bells spray only if fleahoppers reach hazardous levels, and with insecticides that are least deadly to the beneficial insects.

Boll weevils, on the other hand, can usually be managed with a minimum of pesticides if a farmer does his homework. After harvest, the Bells are required by state law to spray, defoliate and cut down the remaining stalks of their cotton plants. If this is done, weevil populations are deprived of food and are unable to go into diapause—a type of hibernation that allows insects to survive the winter. If stalks are destroyed on schedule, few weevils should appear the following season.

Despite its success so far in the Valley, integrated control isn't foolproof. "Weather is the one factor we haven't any control over," says James A. Deer, area entomologist at the Texas A&M research and extension center at Weslaco. "Harvest time is normally our rainy season," he explains, "and when it rains we can't destroy the crops in time to prevent diapause."

Moreover, integrated control is a complicated approach, requiring the supervision of trained entomologists who dispense their advice with the precision of a physician making a diagnosis.

But the biggest hurdle, entomologists agree, has been persuading farmers to gamble with natural controls. "We're concerned about the balance of nature, but the farmer is concerned about the balance of his checkbook," says J. W. Smith, Jr., assistant professor of entomology at Texas A&M. Adds Mr. Adkisson: "Farmers are fearful—if they can protect their investment with a few dollars worth of insecticides, they tend to look at the short-term advantages and ignore the long-term risks."

Actually, the integrated controls being used by most Valley farmers are similar to pest-management techniques practiced by their fathers, before the boom in chemical pesticides.

Just around the corner, however, is an arsenal of exotic new weapons, such as synthetic sex attractants to lure boll weevils into traps, and juvenile hormones that prevent pests from maturing.

By far the most promising developments coming out of research laboratories are new cotton varieties being bred for resistance to insects. Texas A&M scientists have developed new "short-season" cottons that can be harvested and ploughed under early enough to prevent insects from diapausing, so that they can't survive the winter.

Some new strains of cotton are actually toxic to insects. Scientists discovered that certain wild cottons that had grown in Central America for centuries without the aid of man had developed built-in insect resistance. M. J. Lukefahr, research entomologist at the Department of Agriculture laboratory in Brownsville, Texas, has been cross-breeding these wild varieties to transmit their resistance characteristics to commercial cotton.

He has come up with new varieties that produce a 75% reduction in each generation of pest—close to the kill rate offered by insecticides. Under ordinary conditions, yields are somewhat lower than commercially grown cotton, he says, but in areas where the budworm is out of control, the new strains "significantly outyield" nonresistant commercial varieties.

His new cottons won't be available commercially for several years, Mr. Lukefahr says, because of the painstaking process of developing resistant varieties that also have suitable fiber properties. However, he says, "There's no reason to think that eventually the new resistant strains won't be completely comparable."

Once these cottons are available, new varieties will have to be developed constantly, he says, if man is to outwit the insect. "Pests develop resistance to plants, too," he sighs. "We'd be naive to think any single variety will offer permanent insect resistance."

—ELLEN GRAHAM

Sensitive Subject

Addressing a group of cattlemen in Chicago in the spring of 1974, Ellen Zawel, a prime mover in the 1973 consumer beef boycotts, received a surprisingly cordial reception until she slipped into her speech one request: "Please don't put that cancer-causing agent back in our meat."

Her reception turned decidedly frigid at that point, because Mrs. Zawel had introduced a sensitive subject among cattlemen these days: the growth stimulant diethylstilbestrol, better known as DES, which is used by cattle feeders to fatten their animals faster and cheaper.

Though a ban on DES has been overturned, the synthetic hormone remains in a legal limbo that leaves cattlemen uncertain whether to use DES and unable to get it if they want to. The DES shortage will contribute to reduced beef supplies and higher prices in the future, cattlemen say.

DES, which enables a steer to gain 10% to 15% more weight in a 230-day period than it would without the hormone, was banned by the Food and Drug Administration in 1973 after tests apparently showed cancer-causing residues of the substance in the livers of some animals. That ban was overturned on Jan. 24, 1974, by a federal court, which ruled it illegal because the FDA didn't conduct public hearings on the matter.

Since the ruling, however, only one DES maker has decided to resume production; the others are waiting until the legal status of DES is clarified. That might take a while, however.

In the meantime, DES will continue to be scarce

and expensive—when available, it costs nearly double its price before the ban.

Many cattlemen say they wouldn't use the drug even if it were available. "Quite frankly, we're just a little scared of what they (the FDA) might do," says Doug Florence, general manager of Stratford of Texas Inc., a computerized 185,000-head cattle-feeding operation in the Texas panhandle. "I'd hate to have the government quarantine my entire herd just because they had been dosed with DES." Adds a spokesman for the Texas Cattle Feeders Association, which represents Stratford and most of the feedlots in Texas: "Our people are being very cautious these days. We don't want to do anything to stir up consumer rancor like we had during the boycotts."

Yet, despite such fears, some cattlemen have been using DES whenever they can get their hands on it. Nearly all existing supplies, however, are meant to be implanted in the steers, not used in the feed. By and large these implants, which were in stock before the ban, don't match the results of the type used in feed, and also increase labor costs, cattle feeders say.

Other than consumer groups, about the only people that aren't complaining about the curtailed use of DES are the makers of DES substitutes, used as implants, whose safety hasn't been challenged. New York-based Commercial Solvents Corp. says sales of its stimulant, Ralgro, increased "fivefold" during the DES ban. Syntex Laboratories, based in Palo Alto, Calif., which makes a stimulant called Synovex, says it has had a "substantial" sales jump, a spokesman says.

But Ralgro and Synovex implants are even more expensive than DES and are generally less effective, some cattlemen say. Duane Flack, vice president and resident veterinarian of Monfort of Colorado Inc., which feeds 200,000 cattle in Greeley, Colo., is among those who prefer DES. Dr. Flack contends the loss of DES contributed to Monfort's $3.7 million loss in the six months ended March 2, 1974.

Monfort's story isn't all that unusual, however. The loss of DES was "a real tragedy" for the entire industry, says Charles Pratt, general manager of the National Livestock Producers Association, the nation's largest livestock-marketing cooperative. Mr. Pratt says cattlemen's costs increased at least $629 million during the ban period because about 17 million head of feedlot-fed cattle slaughtered then each cost about $37 more to fatten than they would have had DES been used.

Dawe's Laboratories Inc., Chicago, is the only one of three major U.S. DES makers that has committed itself to producing the drug. "It's an investment risk," concedes Robert Huelsebusch, vice president, marketing. But if DES is given the green light, Dawe's will have a good chance of taking over the lion's share of the market, he says. Elanco Co., a division of Eli Lilly & Co., has been the leading producer. Chemetron Corp., Chicago, is the other major U.S. maker.

Meanwhile, the FDA has published in the Federal Register a proposal to revoke the only two presently approved methods of identifying and measuring DES residues in meat because they failed "to meet the necessary requirements of accuracy and sensitivity."

DES makers charge the proposal is a maneuver designed to ban DES effectively, if not officially, because tests for DES in meat are required and the FDA would have those tests outlawed.

Cattlemen remain unconvinced and are urging hearings that would consider all the evidence about the alleged danger of DES. DES users and producers say it isn't consumed in meat in large enough amounts to be harmful.

—GEORGE GETSCHOW

Cleanup Campaign

Officials at Ohio Feed Lot Inc., in South Charleston, Ohio, say they deal profitably in beef and purified animal by-product. Others would say, with some justification, that Ohio Feed Lot Inc. deals profitably in euphemism, since the company's "purified animal by-product" is basically manure. But whatever the nomenclature, the company's business represents what might be called the ultimate response to cleaning up the environment.

It works this way: Ohio Feed Lot has about 16,000 head of beef cattle housed in eight metal barns, each barn being more than a quarter of a mile long. Pens are bedded with a few inches of ground bark, sawdust, and other wastes obtained free from wood-products plants. Specially equipped dump trucks drive through the barns, automatically dispensing a computer-selected ration of corn, soybeans, dehydrated alfalfa and various minerals and vitamins.

Just putting the cattle under roofs is a departure from most cattle feedlots. But another new twist adopted by Ohio Feed Lot takes place every two or three weeks, when big tractor-mounted loaders clean out the pens. The mixture of manure and bedding is trucked to another metal building and dumped into a huge concrete trough, 120 feet wide, 700 feet long and 15 feet deep. A system of fans and ducts blows air through the material, assisting bacteria in breaking it down. After five to seven days of periodic stirring, what began as a distinctly unpurified by-product is completely pasteurized, deodorized and ready to be ground, automatically packaged and sold in 50-pound plastic bags as garden fertilizer.

"The bacteria do the work," says John Sawyer, Ohio Feed Lot's president. "We provide the environment." And reap the profits, he might add. The 50-pound bags retail from $1.59 to $2.49, and while Mr. Sawyer declines to discuss specifics of his privately held concern's financial results, he will say that his fertilizer operation is profitable, even "sometimes more profitable than the beef."

Mr. Sawyer's operation is highly unusual in an industry that has traditionally treated animal waste as an unpleasant by-product to be ignored whenever possible. As one result, the areas surrounding large cattle farms have been subject to odors. Worse yet, rain falling on the pens carries manure into streams and ponds. Consequently, tightening restrictions on air, water and land pollution have begun to shut down a handful of traditional open-air beef feedlots.

"There's no question but that a lot of the livestock industry is ultimately going to have to go to enclosed systems," says E. Paul Taiganides, professor of environmental engineering at Ohio State University. He says that "environmental regulations have prevented new operations from starting and many others from expanding."

Many farmers agree that eventually they will have their livestock and some type of manure-processing plant indoors. Indeed, some besides Ohio Feed Lot have already done so. A few hog farmers, for example, are using a water and pump system to carry animal waste away from the pens to a treatment center for eventual use as fertilizer. However, for most livestock farmers, the capital costs required for an indoor manure-treatment operation—billions of dollars nationwide—will defer such changeovers to some years in the future.

Meanwhile, many face immediate demands that they reduce pollution from the huge quantity—more than a billion tons—of manure that their animals produce each year.

The first pressure from the new environmental regulation is to eliminate water pollution. According to pollution authorities, a 50-acre feedlot on a rainy day can produce runoff into neighboring streams that is equivalent to the raw sewage from a city of 60,000 people. Therefore, many feedlots are being forced to install systems to prevent fouling the waterways. As one example, Webster Feedlots Inc. of Greeley, Colo., recently completed, for about $30,000, a waste-retention system whereby manure-laden water is collected in ponds and then is pumped out to irrigate and fertilize a green belt of grass and trees ringing the operation.

The odor problem is more difficult for the livestock farmers. Some researchers foresee the day when a controlled diet will result in odorless manure; but until that time farmers are being forced to learn some rudimentary public relations—such as immediately plowing manure under the ground—in order to forestall complaints from neighboring residential areas.

For some feedlots, the simplest answer has been to move their animals away from settled areas. Monfort of Colorado, one of the world's largest cattle feeders, is spending about $6 million to move one of its 100,000-head feedlots a dozen miles away from the city of Greeley. "The feedlot has been here since 1930, and the city has grown out toward it," says Duane Flack, general manager of Monfort's feedlot division. "We aren't under legal pressure to move, but we could see it coming."

Other feeders are fleeing the corn belt for the semiarid West, where there is lower population density and where lower rainfall minimizes odor and polluting runoff. In the process, however, the cattle are moved away from the cheapest grain and the major meat markets.

The long-range answer would appear to be a system like that of Ohio Feed Lot Inc. And if the costs of converting to such a system are high, some in the industry figure that these costs will be more than offset by profits derived from the manure. Ohio Feed Lot sells its purified manure for fertilizer. But some researchers feel that manure will soon be even more valuable as a source of energy or feed.

There's nothing new about using manure for energy. Pioneers burned dried bison dung, which they dubbed buffalo chips, to heat their sod shanties. In this century, methane from manure has been used for power in European farm hamlets when natural gas was hard to get. And while the costs of constructing plants to produce energy from manure on a large-scale basis would be high, some in the energy field believe that a prolonged fuel shortage would make such plants economical.

Even more promising is the potential of purified manure for use as feed. Organic Pollution Control Corp., a Grand Haven, Mich., subsidiary of Technology Inc., says its $100,000 dehydrator can turn poultry manure into cattle feed at a profit. The company claims that costs run about $35 a ton of dried manure, which is then worth about $100 or $120 a ton as a protein supplement. Methane from the manure provides part of the required energy, but the machines do need some additional natural or propane gas or fuel oil.

Ceres Land Co., in Sterling, Colo., is also converting manure—but from cattle, not poultry—into feed. The process involves leaching out protein and fermenting the remainder. The end result, Ceres says, is high-quality protein and a roughage comparable in flavor and food value to corn silage. What's more, the company says, the resulting feed costs only a third as much as an equivalent amount of grain, and the only wastes are salts and ash, roughly 5% of the total.

Further experimentation and Food and Drug Administration approval will be required before the use of manure-feeds becomes widespread; but researchers say that eventually wastes may supply up to 20% of the nation's livestock feed, freeing more grain for human consumption. Ceres is completing a facility that will handle manure from 10,000 beef animals, and the company has plans for even bigger units that would process waste material from 200,000 cattle.

—RALPH E. WINTER

Part Six

AGRIBUSINESS

The farmer is surrounded by a corporate world known as agribusiness. He buys his machinery and supplies from these concerns, and he often sells his crops and livestock to them. Sometimes the "them" are cooperatives that are owned by groups of farmers, but these operate very similarly to big companies. Many farmers resent and distrust the giant corporations that they must deal with—especially when companies try to compete with them in producing food and fiber. But agribusiness is what provides farmers with the high-technology inputs they need to keep boosting production, and agribusiness is what links the farmers to their customers across the country and around the world.

Agribusiness Boom

On a sunny, late-summer day in Morris, Ill., Earl Ripsch hurries outside his office to help direct traffic in the small farm town. Trucks hauling soybeans to his elevator are backed up to the courthouse four blocks away. Mr. Ripsch manages the elevator for Illinois Grain Corp., a farmer-owned cooperative that made more money in its fiscal 1973 than ever before.

A hundred miles to the west, in Moline, Ill., William A. Hewitt glances at the primitive African art decorating his plush office and ponders a statistic: The company he heads, Deere & Co., sold $1 billion of tractors and other farm equipment in the first nine months of fiscal 1973—as much as in all of fiscal 1972.

In Sioux City, Iowa, an outraged James McDonald fires off a telegram to President Nixon complaining of price ceilings. "We are completely dumbfounded. . . . It is a tremendous injustice," he writes. For Mr. McDonald's company, Flavorland Industries Inc., and for most other meat packers, the 1973 farm boom was frustrating.

In his Bakersfield, Calif., office, 20 miles from his company's nearest crop, Fred Andrew studies a computer projection of Superior Farming Corp.'s operations 20 years from now. Mr. Andrew dreams of greenhouses on skyscrapers, of entire farms under plastic and—perhaps most fanciful of all—of making a lot of money in corporate farming.

Such is the breadth and diversity of U.S. agribusiness—a $140 billion network of companies and cooperatives that supplies the farmer with what he needs to produce crops and livestock, that processes and distributes farm products, and that in some cases tries to raise them. Agribusiness' fortunes rise and fall with the farmers', and in 1973 for the most part they were marching to the bank together.

Deere & Co.'s earnings in fiscal 1973 jumped 50% from a year earlier. Earnings of International Harvester Co., another farm-equipment maker, climbed 23% during the same period. International Minerals & Chemical Corp., a fertilizer producer, saw its fiscal 1973 earnings rise 28%, while its profit in the first quarter of fiscal 1974 increased fourfold. Among processors of farmers' products, Central Soya's profits soared more than 60% and Archer-Daniels-Midland Co.'s climbed 42% in fiscal 1973. At DeKalb AgResearch Inc., a seed producer, earnings rose by a third in its first six months, and Northrup, King & Co., another seed company, posted a whopping 95% gain in its first nine months. Monsanto Co. earnings, a substantial portion of which come from sales of farm chemicals, jumped to $6.90 a share in 1973 from $3.49 in 1972.

All this didn't go unnoticed on Wall Street. After shunning agribusiness companies for years because of erratic earnings and general lack of glamour, investors chased agribusiness stocks as though, as one analyst puts it, "the farm belt is the new IBM."

Wall Street's enthusiasm was even surpassed in many boardrooms. "We're witnessing the early stages of a period of greater sustained agricultural growth than any we have witnessed since World War II," said Mr. Hewitt, chairman of Deere.

But that isn't how it looked to Mr. McDonald of Flavorland and other meat packers, who say their hopes for tidy profits were dashed by the government. After riding out the consumer boycott and price controls early in 1973, packers as late as May remained optimistic. Then came price freezes and supply havoc. Many livestock producers held animals off the market, and prices for the ones that were marketed rose above levels that packers could pay and still make a profit from the meat.

Thus, while many other agribusinesses were operating at capacity, 818 of the nation's 3,000 or so packers curtailed operations. Some 170 of those 818 closed entirely, and about 40 of them remained closed until late September, the Agriculture Department said.

A few large packers did manage to increase profits despite the price freeze, primarily by custom slaughtering for supermarket chains. Earnings of Iowa Beef Processors Inc., for example, rose to $4.96 a share in its fiscal 1973 from $3.03 the year before. But most packers "incurred horrendous losses" during the freeze, one industry observer says.

Another segment of agribusiness that isn't doing so well is corporate farming—despite the popular image of big, impersonal companies steadily shoving small-but-honest farmers off the land. In 1969, an Agriculture Department study showed, corporations with 10 or more shareholders owned only 0.08% of all U.S. farms, accounting for just 2.9% of total farm sales. Since then, corporate farming has shrunk to even less significance as Gates Rubber Co., S. S. Pierce Co. and CBK Agronomics Inc., among others, pulled out of highly publicized farming ventures.

Those that remain are losing money and retrenching. United Brands Co.'s lettuce-growing and floriculture subsidiaries lost $2.5 million in 1972. Tenneco Inc. has cut back its farmland by 70,000 acres and farms only 20,000 acres of specialty crops, while trying to build a brand-name market for produce grown by others for the company. Purex Corp. has cut down to 10,000 acres of vegetables from 40,000, and it intends to abandon vegetables entirely in favor of crops "where there

isn't a high labor content," says Louis Byington, treasurer of the bleach and detergent maker. He declines to
say how much Purex has lost on farming since it started
in 1968.

"We don't go around publicizing this aspect of our
corporate life," he says. "But, oh yes, we lost a lot."

Purex blames its failure on an expensive labor contract with the United Farm Workers Union. Most of the
big corporate farms are in California and other Western
states because the mild climate is conducive to a variety
of crops and year-round operations; but militant farm
unionists help make the labor situation in the West
anything but mild.

Labor troubles aren't corporate farming's only
problem, though. Companies' "financially oriented
brass didn't really understand farming," concluded
Farm Journal, a trade magazine, in a recent survey. For
one thing, the executives were used to steady prices, not
the roller-coaster swings common to commodities. For
another, they hired some free-spending farm "business
managers" who lacked the family farmer's sense of
thrift and willingness to work long hours.

"What man's going to leave his warm bed on a cold
winter night to go sit up with the company sow?" asks
Agriculture Secretary Earl L. Butz, metaphorically dis-
missing the "threat" of corporate farming.

Fred Andrew is still trying to make a go of corporate farming, though. As president of Superior Oil Co.'s
Superior Farming Corp., he oversees what is probably
the most technologically advanced farming operation in
the country.

Superior Farming employs about 1,200 people to
raise 27 crops on 35,000 acres in California and Arizona.
It uses such advanced practices as a computer-pro-
grammed and monitored "dripping" irrigation system
in which water trickles to roots through taps installed

at each plant's base. In Tucson, the company has a 13-acre, plastic-enclosed "tomato factory" in which every agronomic variable is strictly controlled. Mr. Andrew says the operation produces eight times more tomatoes than comparably sized field operations.

Such innovations cost money, of course, "but we don't believe in going the cheap route," he says. "We're more interested in the least-cost method over a long period." These outlays and the overhead of large technical and management staffs have kept Superior Farming in the red since it started in 1968, but Mr. Andrew predicts that "we'll turn the corner" before long.

Arizona-Colorado Land & Cattle Co. is already making money by taking a broader approach to agribusiness. In 1965, Michael Geddes and several others bought the ailing company mainly for the 400,000 acres of land it owned. The company since has expanded its holdings to 1.2 million acres, which eventually will be developed into shopping centers, residential areas and so on.

Meanwhile, Mr. Geddes is using the land as a base for its agribusiness operations. From cattle ranching, which it got into "to help pay the mortgage," the company expanded into cattle feedlot operations and meatpacking. It also has acquired a thriving specialty equipment manufacturer and a commodity-futures brokerage firm.

The strategy has paid off handsomely. Sales have risen from $15 million in 1968, when the company went public, to $128 million in 1973, and profit zoomed from $800,000, or 29 cents a share, to $5.7 million, or $1.55 a share. Agribusiness contributes about half of the pretax earnings, and land sales and development the other half.

"There's money to be made in agribusiness," says Mr. Geddes, a Harvard Business School graduate who at age 34 is president of the Phoenix-based company. It is one of the five largest cattle ranchers in the country, but, Mr. Geddes says, "our cattle ranching has contributed practically nothing to our profit growth."

In some segments of agribusiness, corporations are far overshadowed by the 7,700 farmer-owned cooperatives. Most of them are small; more than 60% have sales of less than $1 million. But at least six have shouldered their way onto the Fortune 500 list, and the biggest, Associated Milk Producers Inc. in San Antonio, has sales exceeding $1 billion. Together, the co-ops account for an estimated $25 billion in sales—about 17% of all agribusiness volume.

And the volume is growing. The Agriculture Department says that co-ops now sell 26% of all commodities marketed in the U.S., up from 20% in the early 1950s. That includes 73% of all dairy products, up from 53%. Co-ops also provide 16% of all the feed, fertilizer and other supplies used by farmers, up from 12%. "We don't apologize for our growth," says E. V. Stevenson, executive vice president of FS Services Inc., a large supply co-op based in Bloomington, Ill. "Maybe we're growing because we're doing a better job" than corporations.

FS Services earned a record $16.4 million in the year ended Aug. 31, 1973, on sales of $269.5 million, also a record. It began in the 1920s when some farmers decided to pool their dollars to buy kerosene, and it now numbers among its properties six feed mills, three fertilizer plants, three petroleum terminals and four seed-processing plants. It also owns sizable pieces of CF Industries Inc., a Chicago-based cooperative that is one of the country's largest fertilizer suppliers, and Illinois Grain Corp., a grain-marketing cooperative that has the same directors as FS Services has.

Because co-ops generally aren't taxed on their earnings if they return them as "patronage refunds" to members, many corporate executives still view co-ops as "quasi-socialistic entities presenting unfair competition," says Ray A. Goldberg, professor of agribusiness at Harvard Business School. That's changing, though, he says.

"If processing and distributing corporations have learned anything, it's that supplies of the commodities they handle can no longer be taken for granted. And producers of commodities learned that their prosperity depends largely on reaching international markets, which multinational corporations can do much better than they."

The upshot, Prof. Goldberg says, is an increasing number of joint co-op-corporate ventures in production and marketing of certain commodities. CPC International, W. R. Grace and Ralston Purina are among the companies that already have adopted this strategy, he says.

Meanwhile, agribusiness is the subject of a stepped-up antitrust campaign by the government. In testimony before a House antitrust subcommittee in June 1973, Thomas E. Kauper, assistant attorney general, pledged "to expend a substantial degree of our resources" probing antitrust violations in agribusiness. Sheer size is a major target of the trustbusters, and the co-ops' tax exemption may be another. Several probes are under way, and some challenges to acquisitions have been filed.

But agribusiness executives aren't letting such concerns dampen their enthusiasm. "All things considered," says Mr. Stevenson of FS Services, "I'd have to say prospects are the brightest they've ever been."

—JOSEPH M. WINSKI

Firms as Farms

The man behind the big desk was wearing a large grin and a red-and-white checkerboard necktie. He had just come to work for the company, but already he was earning a salary "in the high five figures." The job suited him fine, he said. "I'm dealing with problems of nutrition world-wide. It's a real challenge."

The man was Clifford M. Hardin, who resigned in November 1971 as U.S. Secretary of Agriculture to become vice chairman of Ralston Purina Co., whose corporate symbol is the red-and-white checkerboard. His successor as Agriculture Secretary, Earl L. Butz, is a former Ralston Purina director.

Some people in agriculture weren't nearly as pleased with Mr. Hardin's corporate post as he was. They regarded his move and his replacement by Mr. Butz as evidence of the growing "takeover" of the nation's agricultural economy by concerns like Ralston Purina—huge, highly integrated "agricorporations" that do everything from raising crops and animals to running restaurants.

The controversy over such companies isn't new, but it has been more intense of late as farmers strain to increase their incomes to match the rising prices of what they buy. To its critics, the agricorporation has come to embody forces pushing farmers off their land and making the lives of those who remain more difficult by holding down the prices of their goods. To the corporations' defenders, the firms are working mightily to improve the nation's standard of living and are providing consumers—including farmers—with better foods more cheaply.

The issue has strong political overtones. Some farmers' groups, notably the National Farmers Union and the National Farmers Organization, charge that the agricorporations use their influence in Washington to ensure bumper crops of grain and livestock, reducing farmers' income per unit of output while cutting the raw-material costs of the big food and feed processors.

The cry has been picked up by some Democratic presidential candidates. Sen. George McGovern of South Dakota drew applause from an Illinois farm audience during the 1972 campaign when he warned against being "conned by the disastrous defenders of the Nixon, Hardin, Butz, Ralston Purina . . . corporate line on agriculture."

Like other companies in the field, Ralston Purina denies hewing to any "corporate line" on farm production. "That's just political rhetoric," says R. Hal Dean, the concern's chairman. "They call us a giant agribusiness as if there were something dirty about it, but we're very proud of our partnership with farmers and our contributions to the nation's food supply."

But company officials don't deny that the mass processing and marketing techniques, requiring a large and steady flow of farm goods of stable price and quality, have done much to change the face of American farming over the past several decades. "Farming isn't a way of life anymore, it's a way of making a living," says William T. Lane, a Ralston Purina vice president. "That's a fact, whether or not you like the sound of it."

"Agricorporation" is a loose term, covering a sizable number of companies that engage in different aspects of raising, processing and selling food products. Some concerns, such as Swift & Co., Stokely-Van Camp Inc. and Consolidated Foods Corp., are primarily in the food business while others, like Dow Chemical Corp. and several big oil companies, engage in agriculture as a sideline to their concentration in other industries.

Yet perhaps none fits the agricorporation mold as fully as Ralston Purina. The company was founded in St. Louis in 1894 as Robinson-Danforth Commission Co. by George Robinson and William H. Danforth. The two had observed that horses and mules frequently developed colic when fed a diet of straight corn, but they found that oats, the alternate feed, was too expensive. So they ground and mixed the two grains into a feed that was cheaper than oats and safer than corn.

In those days, the company wasn't much different from thousands of other local feed companies, but Mr. Danforth had a flare for promotion and set out to develop a brand-recognition program for his product. He adopted the checkerboard trademark from a neighborhood farmwife who dressed her children in red-and-white checks so she could spot them in a crowd.

The company started making a breakfast cereal four years later, calling it "Purina" to denote purity. The firm changed its name to Ralston Purina in 1902 when Mr. Danforth persuaded a Mr. Ralston, who ran a popular health club, to endorse the cereal. Mr. Ralston, in return, asked that the company bear his name.

Ralston Purina concentrated on feeds and cereals until the 1950s, when it began selling pet foods. It went into poultry production in 1961 (but sold its broiler-chicken operations in 1972 because it was losing money), the sale of seafood products (Chicken of the Sea) in 1962 and fast-food restaurants (Jack-in-the-Box) in 1968. It also leases breeding stock to hog producers and extends credit to farmers who buy its feeds. Sales in the fiscal year ended Sept. 30, 1973, were a record $2.43 billion and profit was a record $77.5 million, or $2.22 a share.

As a result of this diversification, feed sales now

provide 52% of Ralston Purina's revenues, down from about 90% in the late 1940s. Still, it is in the feed area that the company exerts its broadest influence.

In terms of sales, it accounts for 14% of the U.S. livestock-feed market, a proportion larger than the combined total of its four biggest competitors. To fuel its 90 feed mills around the country, the company buys more than 1% of the nation's total annual output of corn, wheat, milo and soybeans, making it the largest single domestic consumer of grains.

Ralston Purina's impact on the rural economy might best be illustrated by its effect on the life of a hypothetical American farmer. Before World War II, the typical U.S. farmer kept a few head of cattle, hogs, poultry and draft aimals, and he grew grain to feed them. The farmer's wife kept a vegetable garden. Cash income came from selling what the family or its livestock didn't consume.

Today the typical farmer grows only grains. Instead of gardening, his wife has a job in town. He sells his grains to Ralston Purina, which turns them into animal feeds. The resulting animal products—meat, dairy foods and eggs—turn up in his local supermarket. Instead of eating bacon and eggs produced on his own farm, he breakfasts on the store-bought varieties raised on Ralston Purina feeds or he eats Ralston Purina Wheat Chex cereal. For lunch, he has a Chicken of the Sea tuna sandwich. There won't be as many scraps for his dog, but that's okay, because he has bought Purina Dog Chow. For a meal out, he treats his family to dinner at a Jack-in-the-Box restaurant.

Whether these changes are for the better is in dispute. Obviously, fewer farmers are around now than 40 years ago, and the number is expected to shrink further in the years ahead. But the survivors, experts agree, will be large enough to deal effectively with their corporate suppliers and customers.

Take Ralston Purina's role as a buyer of farm commodities, for instance. Because of its size and the nationwide distribution of its facilities, it is able to achieve market benefits that don't accrue to smaller buyers. It buys grains from independently owned elevators scattered around the country, but its purchasing is controlled from its St. Louis headquarters, where specialists daily consider price and supply data from its local agents. With the help of computers, they determine the least expensive ingredient mix for their feeds that will meet nutritional standards.

The price of grains is influenced by national and international factors, but Ralston Purina still has a good deal of leverage. When the price of one grain rises, local buyers are instructed to switch their purchases to cheaper substitutes or to buy in nearby regions where prices are lower. The net effect, of course, is to hold down the per-bushel prices that farmers and middlemen can obtain. A large farmer can increase his production to compensate for this loss, but the smaller farmer doesn't have such an option.

Ralston Purina executives say their methods of purchasing farm products aren't any different from those of big users of any other kind of raw material; they believe that no onus should be attached to them. They go on to say that whatever negative impact they may have had on farmers has been more than offset by the benefits that livestock and poultry producers have derived from the use of their feeds.

Indeed, it is widely agreed that advances in livestock-production efficiency have been dramatic in recent years, owing greatly to research by Ralston Purina and others in the industry. Today an animal will consume less feed than previously and grow larger and faster.

For example, Ralston Purina men say that to gain 100 pounds, a hog used to have to eat 12 bushels of corn weighing 680 pounds. The same weight gain now can be achieved by feeding it five bushels of corn weighing 285 pounds plus 45 pounds of the company's "hog chow." What's more, it takes just 4 months to fatten a hog for market now, down from eight months in 1930.

Similar gains have been achieved in raising other types of meat animals, company officials say.

Besides selling feed, Ralston Purina also gives technical advice to customers. It says that about 1,000 of its 5,200 dealers are specialists in livestock management and that those who aren't can refer customer queries to the company's laboratories. Livestock feeders send samples of their pasture grass to Ralston Purina for analysis, enabling the company to recommend supplementary feeding programs.

But agricorporations—not including Ralston Purina—have made their presence felt in the way livestock has come to be purchased, and there have been casualties among small feeders. The meat-packing industry has left the big cities and scattered regional plants near sources of supply. Now, instead of bidding against one another at the Chicago stockyards, buyers for corporations such as Swift and Armour & Co. can switch their purchases back and forth among various suppliers to obtain meat more cheaply.

Agricorporations also are making their presence felt on the farm scene by producing their own crops and livestock; according to the U.S. Department of Agriculture, about 100 publicly held companies engage in such pursuits.

Ralston Purina has been active here, but its major experience as a direct producer wasn't a success. In 1961 it went into the production of broiler chickens as part of a general corporate move and became a major force in the industry.

The application of corporate methods of raising chickens has reduced costs by some 25% and tripled per-capita consumption of the meat since 1950, but it also led to overproduction and sagging prices that affected large producers as well as small ones. So, in the spring of 1972, Ralston Purina sold all its broiler facilities because of low returns. Within less than two years, Pillsbury Co. had followed suit.

In the long run, however, the agricorporation's influence on American agriculture seems sure to grow, and the individual farmers who survive will be the ones who can use technology to expand their output and income.

"Farmers just can't be separated from the agribusiness system anymore," says Ray A. Goldberg, professor of business and agriculture at Harvard Business School. "The ones that have the flexibility to respond to it and the consumer demands behind it will come out best."

—FRANCIS L. PARTSCH

The Chicken Czar

"It takes a tough man," says the Chesapeake-accented voice of Frank Perdue on television, "to make a tender chicken. . . . If you want to eat as good as my chickens, eat my chickens."

Millions of people — in New York, Maine, New Hampshire, Massachusetts, Rhode Island, Connecticut, New Jersey, Pennsylvania, Ohio, Delaware, Maryland, Virginia and Washington, D.C.—are doing just that. For the several hundred years up until 1971, most Americans ate their 41 pounds or so of chicken each year knowing only that it came from an uninteresting and ungainly barnyard bird of obscure origin. Then Mr. Perdue managed his improbable feat. He turned a seemingly homogenous agricultural product into a brand-name delicacy and, in the process, turned his family's chicken and soybean ventures into a business with sales of $120 million a year.

How did he do it? Experts say by working hard, by feeding his birds stuff that turns them yellow and by putting his own name on the line. In an age when suspicious consumers are subjected to a steady diet of TV commercials showing dubious charts, cutesy man-on-the-street testimonials and staged tests, Mr. Perdue simply and openly sells himself. It's the head of the company in those commercials, asking for your gripes, promising your money back if you don't like his chickens. "Add to that a face that's so plain it's got to be honest and," an ad executive says, "you can't help but believe in the guy and his product."

Privately held, Perdue Inc. ranks among the top 15 chicken packers in the country. More Perdue chickens than any others are sold in the big New York market. And, because Mr. Perdue advertises his chicken as a top-of-the-line product sold only in the highest-quality markets, he gets a better price than most of his competitors.

Now other big regional chicken packers, including Holly Farms of North Carolina, probably the biggest, are branding their products. Not all chicken men like this trend, and some of them are crying foul. "All broilers are produced in the same way," says an official of the National Broiler Council, an industry group based in Washington. "But Frank's got them convinced in New York that his are better because they're more yellow." Perdue chickens are yellower, he says, because of the volume of xanthophyll, a chemical naturally occurring in corn, egg yolk and marigold petals, that goes into their feed. "In the South," he says, "people actually prefer a much whiter bird than a Perdue chicken."

That makes Mr. Perdue see red. "It costs me $1 million a year to give my chickens their healthy yellow color," he says. "If I'm going to spend that much money to get a bird that looks healthy, you know I'm going to bust my butt to make it healthy, too."

Indeed, many shoppers sing high praises to Perdue quality. "I always have to poke a hole in the package of another brand of chicken to make sure it doesn't smell bad," says Mrs. Helen Nabstedt, of Montvale, N.J. Mrs. John Dorsett of nearby Park Ridge says: "I once cooked a couple of chickens with another brand and, boy, did they smell and taste bad! I've never had a problem with Perdue."

Perdue's two processing plants in Maryland and Virginia, employing most of the company's 1,800 workers, can pluck and pack more than 300,000 birds a day, or 78 million a year. (A third plant is going up in North Carolina.) In a good year, Perdue can earn a nickel or so profit a pound from chickens. In a bad year, the firm could lose a few cents on every bird.

Though Mr. Perdue won't tell his firm's profit, he says he never has finished a year in the red since he joined his father, Arthur Perdue, in the chicken business in 1939. Arthur Perdue started the business in 1920 with 50 hens that cost him $5 and a chicken coop that he built himself.

Frank Perdue made the business scientific. He used computers to figure the cheapest way to turn the least feed into the most meat. He hired geneticists to breed bigger-breasted birds and veterinarians to produce healthier ones. Still, he says, "This is the kind of business where a Harvard MBA is out in left field. My advantage is that I grew up having to know my business in every detail. I dug cesspools, made coops and cleaned them out. I know I'm not very smart, at least from the standpoint of pure I.Q., and that gave me one prime ingredient of success—fear. I mean, a man should have enough fear so that he's always second-guessing himself."

Perdue Inc. prospered modestly for years. Then in 1968, Frank Perdue started branding his chickens instead of merely wholesaling them anonymously. By July 1971, he took the big plunge and started consumer advertising in New York. Since then, his yearly ad budget has grown to about $1 million from a few hundred thousand dollars.

Perdue normally earns about 75% of its profit from dressed broilers; the rest comes from ancillary services and byproducts, such as hatching eggs for other chicken producers, and sales of soybean meal, soybean oil and pine-bark mulch, which Mr. Perdue acquires from his chicken-litter supplier and then packages and distributes for gardening use.

Mr. Perdue spends a good part of his time in

friendly persuasion. His targets aren't just the cosumers who see him on television, but everybody else involved in the production and marketing of his chickens —including the growers who raise Perdue-hatched and Perdue-processed birds, the distributors who wholesale them and the retailers to whom Perdue sells directly.

One morning in the spring of 1974 finds Mr. Perdue in the office of Harold Tarr, executive general manager of Pioneer Food Stores Cooperatives Inc., Carlstadt, N.J. Mr. Perdue welcomes complaints—"I cannot make better chickens if I don't get some beef," one of his commercials says—and this morning he is getting a beef from Mr. Tarr. A Perdue salesman had persuaded Mr. Tarr to feature Perdue chickens in a sale in the 35 Pioneer retail stores that carry meat. So Mr. Tarr revised his scheduled newspaper ads at the last minute to shoehorn the chicken sale in.

But Perdue, Mr. Tarr gripes, couldn't supply enough chickens for the sale. "I had one woman who tore an apron right off our butcher," he says heatedly. "We were the first chain you hooked on with, Frank, I think we deserve better treatment than that."

Mr. Perdue picks up a telephone and calls the salesman. "We're naked on this," he says. "Harold has done us a favor" (by complaining). Later, Mr. Perdue directs his sales force never to commit chickens for a big sale on shorter notice than a week.

"Frank Perdue is the antithesis of the company president," says Donald McCabe, an executive of the Perdue advertising agency, Scali, McCabe, Sloves Inc. "Most guys talk a lot, but who do they talk to? Frank talks to butchers in a Boston ghetto at 5:30 in the morning. He knows the territory and he fights like hell to keep it."

In 1974, Mr. Perdue was slugging it out in the New York market with Pearl Bailey, who pushes, on commercials, the Paramount-brand chicken marketed by Cargill Inc., a big, privately held Midwestern firm. Last year, Mr. Perdue won the "Battle of Boston" for chicken leadership, whipping the Buddy Boy brand of Maryland Chicken Processors. He spent months in Boston lining up major distributors and prestigious retailers. Otherwise, his efforts go mainly toward preventing any defections from his corps of distributors and retailers.

Chicken economics—including shipping costs, and the proximity of feed supplies like soybeans—make it a regional business. Perdue has no plan for going national. And at Mr. Perdue's age (53 in 1974) there's no immediate reason to sell out to the public, either.

Married, with four children, Mr. Perdue lives in a comfortable lakeside home in Salisbury, Md. Outside chickens, his only major interests are tennis, which he rarely has time for, and journalism, which he thinks is biased. On the wall in his office hangs a picture of him playing tennis with former Vice President Spiro Agnew. He employs a cook, hired from his own company kitchen because of her expertise in preparing—yes, that's right.

—BILL PAUL

The Exporters

The companies that export U.S. grain are the shy, retiring types of agribusiness. They avoid publicity, partly because they think nobody understands them, partly because they are very competitive, and partly because most of them are privately owned and thus they normally don't have to court anybody's favor.

So it came as quite a shock to the exporters when they were hauled in like hoodlums to a precinct station and shoved under the bright glare of public wrath and congressional inquiry as a result of the massive grain sales to the Soviet Union in 1972. The public was angry because the exports to Russia set off skyrocketing food prices. Farmers were mad because they had sold much of their grain before prices shot up. Some Congressmen charged that the exporters reaped windfall profits at the expense of taxpayers and consumers.

Testimony at congressional hearings portrayed the export executives as the James Bonds of business. Executives of four companies said they separately and secretly huddled with Russian buyers in a New York hotel suite over the long Fourth of July weekend. They lavishly entertained the Russians in their homes and on private jets and yachts. And, according to detractors, the exporters secretly wangled inside information and financial favors in the form of fat export subsidies from Department of Agriculture officials.

As it turned out, though, the inquiry exonerated the exporters of any wrongdoing, and it indicated that some of them even lost money on the Russian deals. And the exporters themselves insist that their business isn't nearly as clandestine or romantic as the Russian deal suggests.

"More often we're worrying how we can move grain from an elevator in Iowa to the port of New Orleans at an eighth-cent a bushel less than a competitor," says John F. McGrory, an executive of Cargill Inc., an exporter based in Minneapolis.

The controversy over the Russian deal obscures the crucial role that grain exporters play in marketing the output of U.S. farms. Agricultural products are the leading U.S. export, and grains and soybeans are the most important. In the year ended June 30, 1973, farm exports soared 60% to a record $12.9 billion, mostly because grain and soybean shipments doubled to $8.5 billion.

The long-term outlook is bright because rising affluence overseas is accelerating demand for food and feed grains. To make it easier to get at this market, the U.S. is pushing high-level negotiations to persuade other countries to lower trade barriers against U.S. farm products. But the U.S. has no official export agency, such as many countries have, so the commercial exporters are the front-line forces in efforts to build up the farm export trade.

Many grain merchandising and processing firms do some overseas business, but a half-dozen companies account for 90% of U.S. grain exports. The giants are Cargill and New York-based Continental Grain Co. Few U.S. companies have earned more dollars abroad in recent years than these two, each of which is shipping grains at the current rate of about $2 billion a year, trade sources figure. Both also have diversified into feed manufacturing, flour milling and soybean processing. Cargill also is a major producer of salt, and it trades internationally in sugar, metals and ore. It also has gotten into insurance and leasing operations.

The other leading exporters are Cook Industries Inc., based in Memphis, and Bunge Corp., Louis Dreyfus Corp. and Garnac Co., all of New York. Bunge, Dreyfus and Garnac are foreign-owned or controlled. Cook, the only publicly held major exporter, is an old-line cotton merchandiser that diversified into grain 10 years ago.

These firms are doing export business at about half the rate of the top two. In fiscal 1973, Cook's volume of grain and soybeans—most of it for export—soared to $1.5 billion, four times the amount handled five years earlier. Walter C. Klein, president of Bunge, acknowledges that his company moved over $1 billion of grain in 1972.

The big six operate similarly. They all have networks of sales offices and agents throughout the world, working directly for them or for affiliated companies. Most offices are tied to headquarters by teletypewriter systems that pour out thousands of messages around the clock giving details of buy and sell orders, deal proposals, crop conditions, political developments and other bits of intelligence that could affect grain prices. Most exporters have direct telephone lines to import overseas centers such as Paris.

The nerve center is the trading room at headquarters. At rows of desks, a score or more traders conduct virtually all their business by phone—not in hotel suites. Some dealers specialize in a particular grain or oilseed. At one end of the room a large board electronically displays grain prices from the Chicago Board of Trade and other commodity exchanges. In that room or nearby is an open phone line to the exporters' men on the exchange floors.

The traders are on the phones almost constantly, talking to agents abroad about demand or to domestic elevator managers about supply; they always keep one eye on the price board. They can and do trigger multimillion-dollar transactions by uttering a few words. Most have separate phone lines put into their homes so they can take their work home with them.

"Late at night is a good time to call some parts of the world, say like Moscow," says Willard R. Sparks, executive vice president of Cook. A Continental Grain executive claims he once sold $3 million of soybeans from home by phone at 2:30 a.m.

In theory, the exporter's job seems simple: buy grain at the lowest possible price and sell it at the highest possible competitive price. In times of booming demand, such as in 1972-1973, the sale is usually made first. Then the exporter scrambles to find the grain to fill the order. Before the boom, exporters often could buy large quantities of grain from government surplus with minimal effect on the market price. Government stocks are virtually gone now, though, so purchases by exporters to cover significant orders are often accompanied by strong grain-price gains.

The spread between the buying and selling prices must cover several costs. Among them are interest on loans for carrying grain (Cargill and Continental Grain are two of the biggest users of short-term bank credit in the U.S.) and the charges for storing and transporting. The riskiest part is making the sale at the right price, exporters say.

"It's more of an art than a science," declares Melvin H. Middents, a Cargill vice president. "Until a deal is completed we're always wondering if we've guessed right, because if we're wrong we can lose a lot of money."

The exporting game is tremendously complicated in practice, in part because exporters don't deal straightforwardly with customers as a supermarket does. Details of a deal can differ sharply depending on the type of customers—a flour miller, a feed manufacturer or a foreign government agency buying on behalf of a whole country. Sometimes, a customer tells the exporter how much he wants, the exporter quotes a price and a deal is made. Other deals are negotiated, and in still other instances exporters are invited to make offers, or tenders, of grain.

In negotiating a price or making a tender, an exporter has other things to consider—where can he get the grain of the specified quality at a price that will provide a profit after paying costs, and how can he get it shipped on time? The export boom so clogged the seaports and railroads that to help speed shipments and boost profits, the big exporters are acquiring more storage facilities and transportation equipment of their own.

In 1973, for instance, Cargill was planning a large new export elevator at Duluth, while Cook was putting up new elevators in Galveston and Portland, Ore., and Continental was building one at Tacoma. Rumors were that many more port facilities were on the drawing boards.

Cargill increased its fleet of rail cars to 1,500 by the fall of 1973 from 1,000 two years earlier, while Continental tripled its fleet to 1,200. Exporters also were pressing railroads for every available car, to the point that some farm groups complained that rural elevators couldn't get enough cars to ship grain to other customers. Columnist Jack Anderson charged that some exporters made cars available to elevators who would sell them grain for $1 a bushel under the market price. The big exporters denied using such a tactic.

Exporters also added river barges as fast as they could to move grain to the Gulf Coast. Cargill opened its own barge-making factory at Pine Bluff, Ark. in 1973; it has a capacity of 50 barges a year.

After an exporter has made a sale he has to decide

whether to hedge the transaction—that is, purchase futures contracts that give him the right to buy a similar amount of grain later at an established price. The exporter might want to hedge if he has made the arrangement to export but hasn't yet lined up the supplies to fulfill this deal; he especially would want to hedge if he thinks the price of grain is going to rise during this interval. Also, exporters hedge against anticipated price declines when they have large unsold inventories; these hedge sales guarantee them a certain price in the future for their stocks.

Sometimes an exporter might hedge in other commodities (cross-hedging) if price relationships are advantageous. Hedging is such a complicated and sensitive part of the exporting business that most exporters have one executive whose sole job is keeping track of hedging positions.

The need to hedge is one reason exporters say they need to operate secretly. If speculators knew that an exporter had made a big sale but hadn't yet covered with an offsetting hedge, they could jump in the market and drive the price up. The amounts of grain sold to Russia in 1972 were so huge (400 million bushels of wheat, for example) that exporters tried to keep the deals secret as long as possible in order to get covered; the news got out before some of them, including Cargill, had supplies lined up, or had hedged. Exporters often use commission brokers as a cover in trading hedges because if their own men did all the trading they would tip their hands.

"Secrecy worked to the advantage of the Russians, too," recalls Mr. Klein of Bunge. "As it turned out, they used a divide and conquer strategy. None of us were told by the Russians that they were dealing with other exporters. They played us off against each other."

All these decisions and moves must be made within

a strict schedule. Depending on the terms of the sale, either the exporter or the buyer arranges for chartering an ocean vessel, and together they work out a timetable for loading at U.S. ports. The exporter has to coordinate the purchase of grain and the logistics of getting rail cars and barges in and out of port elevators at the right time. Delays in unloading cars or barges or in loading ocean vessels can be expensive.

"Much of the money made in this business comes from holding down the costs of handling grain," says Mr. Middents of Cargill.

To do this, exporters are trying such techniques as shipping grain directly from rural elevators to Gulf ports, bypassing the big, busy terminal elevators. They also are running more "unit trains"—up to 100 cars all loaded with grain that move nonstop to the ports in less than half the time of regular trains.

The exporters make money, but probably not as much as their critics think. Cargill released detailed figures showing that it lost more than $600,000 on its share of the 1972 Russian trade. In 1969, the last time that Cargill let anybody peek at its financial results, the company's earnings of $14 million were less than 1% of its sales of more than $2 billion. That was a low year for exports, however, and trade sources think Cargill's profit margin was higher in the following two years as its sales climbed to an estimated $3 billion.

Cook doesn't break down its earnings from grain. However, E. W. Cook, president, acknowledges that the grain export boom was the main reason the company's fiscal 1973 earnings soared to $23.8 million, or $7.83 a share, from $3.9 million, or $1.26 a share, in fiscal 1972.

—HARLAN S. BYRNE

End of an Era

As the twin-engine Cessna noses down toward a landing, the fields below seem devastated. Yellow weeds choke the earth while only occasionally does an area of lush, blue-green plants appear in neat, cultivated rows.

This is—or was—the land of pineapple—the sweet, tropical fruit that helped nurture Hawaii's economy and became a symbol of Hawaii's fertile life. On the island of Molokai, the 400 or so full-time pineapple workers in the spring of 1974 were anxiously calculating severance pay and other resources for survival. At Maunaloa, the plantation town of the Dole subsidiary of Castle & Cooke Inc., workers were drawing pay for about four days monthly—just enough to till a few fields while the weeds reclaim most of the land. Time hangs heavy, and many men go fishing or gather at illegal but popular cockfights.

Molokai is only the most dramatic and visible sign that pineapple in Hawaii—its second-largest farm crop with a value of about $140 million yearly—is about to take a steep nose dive. Dole and Del Monte Corp., the two leading Hawaiian producers, announced early in 1974 that they would stop all production on Molokai by 1977 on their 17,000 leased acres in an effort to stem heavy losses. No substitute crop is in sight. The close-down, observers say, will lead to a 75% unemployment rate on Molokai, a quiet island of 5,000 people that is untouched by the urban sprawl of tourist influx of the other islands.

Pineapple's problems in Hawaii have ignited a major battle, one that may have an impact on multinational companies everywhere. Critics charge that Dole and Del Monte are prime examples of the dangers of un-regulated multinationals. They point to the acknowl-edged switching of pineapple output by the two compa-nies to areas such as the Philippines and Thailand, where nonunion wages are 10% or less the level of Ha-waii. These critics include the International Longshore-men's and Warehousemen's Union (ILWU), which rep-resents the farm and cannery workers.

The pineapple crisis provokes David McClung, pres-ident of the Hawaii senate and a Democratic candidate for governor, to urge restrictive federal laws and tax penalties for multinational companies. "By what right should our federal laws protect these foreign monsters whose main reason for existence is bigger and bigger profits, lower and lower wages and less and less taxes?" he asked a congressional group rhetorically.

Dole and Del Monte agree that high labor costs—which account for fully half the production costs in the pineapple industry—are a key reason for their decision to cut back in Hawaii and start up plantations and can-neries in the Far East and Africa. But they heatedly deny any suggestion that they are engaged in "runaway production" to low-wage areas. The decision to reduce operations here, they say, came after steady losses since the 1960s and an unsuccessful effort to improve produc-tivity in Hawaii.

Year-round pineapple workers in Hawaii were mak-ing $2.80 or more an hour, sharply above the 15 cents or so in the Philippines, trade sources say. This disparity is partly offset by greater efficiency in Hawaii, but it still costs "more than two times as much to produce a case of canned, sliced pineapple in Hawaii than in almost any other major growing area of the world," says Frank Dillard Jr., manager of Del Monte's Hawaiian division.

Among other costs, he says, is shipping. The Jones Act prohibits use of foreign ships to transport pineapple between Hawaii and the U.S. Mainland. As a result, it costs $40 a ton to ship the fruit in U.S. ships from Hawaii to New York, or almost $10 more than the cost to ship a ton from Taiwan to New York.

Efforts to further reduce the amount of high-cost labor in Hawaiian pineapple growing haven't worked, says Nason Newport, Dole vice president and plantation manager on Molokai. The Cayenne pineapple, the standard variety, has to be planted and harvested by hand, although a mechanical planter is being tested. "We've spent millions looking for a better pineapple than Cayenne, but nothing has been found in 35 years of research," Mr. Newport says.

Whatever happens, the pattern of reduced operations has been set. Molokai's cutback was preceded by the decision in 1973 of Stokely-Van Camp Inc. to close its plantation and the only cannery on the island of Kauai. Dole also announced it was halving its pineapple acreage on Oahu, the main island. In all, six companies have given up pineapple growing here since 1950, leaving only three: Dole, Del Monte and a small, private-label packer and grower, Maui Land & Pineapple Co. Hawaii's total output has been level since the late 1950s, and its share of world pineapple production has sagged to less than one-third from more than two-thirds in 1950.

Nearly all pineapple grown in Hawaii until the early 1970s was canned or turned into juice. But as cannery operations are abandoned, Dole and Del Monte are holding out the promise that a part of the industry here will survive by turning to the marketing of fresh pineapple. They reason that other major producing areas such as the Philippines are too distant to deliver fresh fruit to Mainland U.S. customers. Competitors closer to the Mainland, such as Puerto Rico and Mexico, don't produce fruit that can compete with the quality of Hawaiian fruit, the companies believe.

Initial marketing efforts have been encouraging, both companies report. Dole sold about 42,000 tons of fresh pineapple in the U.S. in 1973, up from nothing five years earlier, says C. M. Waite, senior vice president. That was half the 80,000 tons of fresh pineapple consumed in the U.S. in 1973. Del Monte provided about 15,000 tons of that total, a 70% jump from 1972.

"The marketing effort in fresh pineapple has worked," Mr. Waite says. Most of the output is delivered to West Coast customers by ship, but a 1974 fare reduction granted by the Civil Aeronautics Board in air freight for pineapple makes it economical for the first time to fly the fruit to Eastern and Midwestern cities. "Our fresh operation is profitable," Mr. Waite adds, and he predicts that 200,000 to 400,000 tons of fresh fruit may be marketed by Dole in four to six years.

But even if such dramatic growth is realized, it won't offset the job losses in Hawaii as processed fruit is cut back. Dole processed only 220,000 tons in 1974 in Hawaii, far below the peak of 520,000 tons reached in 1971. And Del Monte by 1976 expects to hire only half the 2,000 seasonal workers for its cannery and field operations that it hired in 1973.

The prospect of high unemployment among pineapple workers on Molokai has led to appeals from the ILWU and some state politicians for sharply higher U.S. tariffs to keep out foreign pineapple. Pineapple from the Philippines and other foreign areas enters the U.S. under a relatively low ad valorem tariff of just over 5%. Many other domestic fruit crops, such as peaches, pears and citrus, enjoy tariff protection ranging from 10% to 35%. But Del Monte and Dole argue against higher tariffs on pineapple, saying it would hurt profits of their foreign operations, which, they say, subsidize losses in Hawaii.

The exact size of those losses—and, indeed, whether they are real—is the subject of bitter dispute. Dole and Del Monte have declined demands from the union to make public their figures on operating losses in Hawaiian pineapple. Mr. Waite of Dole's parent, Castle & Cooke, says only that "it's a lot of money," and Del Monte's Mr. Dillard says that since the mid-1960s the Hawaiian division "has operated at a loss—and the rate of loss has steadily increased in recent years."

Eddie Tangen, an international representative of the ILWU, wonders aloud whether the losses are real or merely accounting gimmickry. And he asks a question of the two big pineapple companies: "How can Maui Land & Pineapple do it when they (Dole and Del Monte) can't?"

His question focuses on a fact that is proving embarrassing to Dole and Del Monte: Maui Land is making a profit from pineapple growing and canning. As a publicly held concern, it publishes its results. In 1973, its profit, largely from 10,000 acres of pineapple, soared to over $1 million on revenue of $28 million; that was about a sixfold earnings gain from 1972. Although income has been slim in prior years, the company has been in the black.

The credibility problems of the big two aren't helped by Colin C. Cameron, Maui Land's president. "I don't know why they aren't making money," he says. But he sees "a lot of room for accounting variations" in big companies. As for Maui Land, he suggests that it has kept its cannery modern and has benefited from growing consumer acceptance of less costly private-label food such as A&P brands rather than nationally promoted brands. Maui sells almost exclusively to outlets such as A&P under private labels. "We're quite optimistic about the future of pineapple" in Hawaii, he adds.

Maui Land's competitors aren't sure why the smaller company makes money. A spokesman for Del Monte suggests that Maui doesn't have the costs of national-brand advertising and sales that Del Monte must pay. But he admits that this is speculative. "It's just a different ball game," he ventures. "You might as well ask Willie Mays why he doesn't hit as many home runs as Hank Aaron."

Maui Land doesn't have any acreage on Molokai, where high-cost water, strong winds and other factors make production unusually expensive. Despite the gloomy economic outlook on Molokai, most of the workers have turned down Dole and Del Monte offers of jobs on the islands of Lanai and Oahu. The companies have offered to pay moving costs, assist with housing and guarantee jobs with no loss of seniority.

"The workers have spent all their lives on Molokai," says Tom Trask, an ILWU representative. "They don't want to move to the rat race of Oahu. They like the (Molokai) life-style." He adds that Dole had planned to nearly triple rents on company-owned housing on Molokai—to $90 a month—after the closedown. "That's how they proposed to soften the blow of leaving, by jacking up rents," he says.

A Dole spokesman flatly denied that the company ever proposed a $90 rent. He adds that the companies have proposed to freeze rents at present levels until the end of 1980 with employes picking up the cost of maintenance that previously was paid by the company. But as of April 1974, ILWU hadn't accepted the offer.

In all the furor over pineapple's decline in Hawaii, few people have been able to see both sides of the issue. One who does is State Sen. Wadsworth Yee, Republican minority leader. "As a businessman, I can understand the companies' problems," he says. "We have the highest-paid agricultural workers in the world, and the companies have to compete against Mainland canned foods like pears and peaches. But it's sad that dollars and cents now mean more (to the companies) than the old Hawaiian spirit."

Mr. Waite of Castle & Cooke replies, "We've been accused of turning our backs on the place that gave us birth. It simply isn't true." He adds that there is nothing sad about facing economic reality "and for the first time we're dealing with reality."

—HERBERT G. LAWSON

Part Seven

RESEARCH FOR THE FUTURE

Much of the spectacular gain in U.S. agricultural productivity in the 20th century has come from improved crop and livestock varieties, better and more knowledgeably used fertilizers and feeds, and more and bigger machines. All these and more are the products of agricultural research, which is continuing to hunt for ways to boost production at the least cost to both farmers and consumers. Much research takes the form of genetic tinkering—a process that often requires years and years to come up with something viable but can, like the introduction of high-yielding hybrid corn in the 1930s, cause widesweeping changes throughout agriculture. Sometimes, whole new crops make their appearance on farms as researchers come up with a way to meet a particular demand.

The Research Farm

Air conditioning improves a hog's sex life.

At least, that's what Swift & Co. researchers concluded during first year testing at the Esmark Inc. unit's research farm near Williamsburg, Iowa.

The farm is one of more than 800 private and government research farms where scientists are trying to find new ways to raise farm products more efficiently. They're looking for breakthroughs in both livestock and crop technology to meet growing demands for even greater productivity from U.S. farmlands.

"There used to be the feeling that if we could produce all we knew how, our troubles would be over," observes W. Robert Parks, past president of the National Association of State Universities and Land Grant Colleges. "Increasingly, that's no longer true. The reserve of unused technology is getting critically smaller, and farmers are coming to the college doors for more information even before we release it through our extension offices."

So the scientists are working harder to find even the smallest improvement in crop and livestock production, because even small improvements can make a big difference, says Mr. Parks. "If farmers could raise one more pig per litter, it would amount to roughly a two months' pork supply for the U.S.," he asserts.

Corporate researchers, who lay out roughly half of the estimated $1.3 billion farm research money spent in the U.S. each year, devote much of their effort to testing products and ideas for their sponsoring companies. But basic agricultural research is a big part of the effort, too, says Glen Heuberger, Swift's dairy and poultry research manager. Mr. Heuberger is one of the designers of the company's $500,000 livestock research farm, which spreads along a ridge in this hilly eastern Iowa countryside.

"Education — passing along what we learn—is as important as our research," asserts Mr. Heuberger, "so we analyze management techniques as well as feed ingredients."

Swift built the Iowa research farm late in 1972, after metropolitan Chicago's sprawl began crowding its Frankfort, Ill., farm built in 1956. Iowa's wide-ranging Midwestern climate allows the testing of ideas and feed ingredients for customers from Georgia to Canada.

"We went to a total-confinement system because that's becoming the trend in agriculture," explains Mr. Heuberger, "and because the nutritional needs we're testing are more critical there. We built on a commercial scale to use standard milled feeds and to test under more realistic conditions." Testing at the farm involves up to 3,000 hogs, 20,000 turkeys and 360 feeder cattle a year.

Some of the early findings appear promising, he continues. Air conditioning the room in which 150 test sows are bred boosted successful matings on first attempts from 50% to above 90%. Keeping the sows in total confinement and feeding them carefully during their pregnancies led to the weaning of two more pigs per litter than the average 7.1 to 7.8 pigs for sows cared for more traditionally. The research farm pigs also tended to grow faster and yielded about 5% more of the main pork cuts.

Feeder cattle, each confined to 20.4 square feet at the Swift farm, gained a pound with one to two less pounds of feed than cows in feedlots five times as large.

And turkeys, confined to four square feet apiece, gained 10% faster with about 5% less death loss than comparable birds raised on open range. More importantly, adds Mr. Heuberger, there are strong indications that confinement growing makes year-round turkey production possible in the upper Midwest, which could lead to doubling production in the area.

For example, Swift had some 10,000 birds in its confinement houses in 1973 when a spring blizzard drifted snow up to the roof tops. More than 250,000 Iowa turkeys on the open range perished in the storm, but none of the confined birds died.

Waste disposal techniques, which are of increasing concern as livestock farms become subject to stiffer pollution-control regulations, are also being tested at the farm. Liquid manure is collected in huge pits beneath the self-cleaning slotted floors on which the hogs and cattle are kept throughout their research farm tests. The liquid manure is odorless until the pits are cleaned, a chore that takes one afternoon two or three times a year.

Paddle wheels, which stir in odor-killing oxygen, are being tested under one of the four confinement hog houses. The oxygen stops the cleaning-time odor, which neighbors around other confinement farms find objectionable, but the wheels cost about 75 cents a day to operate. The Swift researchers are trying to determine if the extra cost is worthwhile.

"The whole confinement idea is an environmental one," Mr. Heuberger says. "You control the animals' environment for more efficient production. And, with planning, you control what you contribute to the environment around the confinement house."

The idea seems to be persuasive. One Canadian hog farmer brought his local zoning officials to the research farm to convince them that liquid manure systems could be pollution-free. The officials subsequently let him double the size of his planned expansion.

While Swift's overall test results are pleasing so far, Mr. Heuberger says that some basic questions remain unanswered. More productive and economical feed formulas are still being sought for all the livestock, for one thing. And while raising beef cattle and pregnant sows in confinement appears promising, the practice has by no means been proven economically viable.

—GENE MEYER

Seeds of History

Even for Louis Bass it's too late to save the Tip-A-Canoe tomato, the pride of the salad bowl back in 1905.

It's gone the way of the Perfection Mountain cucumber of 1933 and the Green Perfection string bean of 1942. Like the dinosaurs, they're all extinct.

The same fate has met hundreds of other plant varieties over the years—from food crops to flowers—as they've fallen from favor with farmers and gardeners. But in an elaborate building at the foot of the Rockies in Fort Collins, Colo., Mr. Bass is putting a stop to this mass disappearance.

As head of the National Seed Storage Laboratory, he has the job of collecting and preserving seeds of every description. As a result, he probably has more seeds than anyone else in the world—hundreds of thousands of them stored in thousands of little tin cans inside 11 huge vaults.

The designers of this million-dollar Fort Knox of the seed world have left nothing to chance. The vaults are fireproof and bomb-proof and are built of 10-inch-thick concrete walls. Inside, the seeds are numbered and catalogued and are kept dormant at a constant temperature of 35 degrees Fahrenheit and a relative humidity of 35%.

Such precautions are necessary, says the U.S. Agriculture Department, which owns the laboratory, because the seeds "are an important natural resource." Their main purpose is as a source of genetic material for plant breeders developing new varieties.

A breeder wanting to develop a larger, better-tasting green bean, for example, might want to use some of the characteristics of the 50-year-old Idaho Refugee green bean. So he would turn to Mr. Bass. In the case of this and several thousand other plant varieties, the laboratory has the only seeds left.

But if a variety has suffered the indignity of neglect and become extinct, the breeder is out of luck. This has already happened with two-thirds of the oats introduced into the U.S. this century and 90% of the soybeans. And several of the 385 new varieties of tomato developed in the U.S. between 1936 and 1969 have vanished without a trace.

The fault lies partly with the breeder himself, says Quentin Jones, an Agriculture Department plant expert. The life expectancy of a new variety, he says, is about five years; then a more bountiful variety is developed "and the old plant is discarded in favor of the new."

The problem of the vanishing plant is even more acute in developing nations, where over the past few years new high-yielding plant varieties have been steadily replacing traditional native crops. Now plant men are finding that cereals, fruits and vegetables that had prospered since antiquity are suddenly hard to find.

So worried is the United Nations Food and Agriculture Organization that it is sending expeditions to such remote areas as Afghanistan and Indonesia to track down native plants. The UN then dispatches samples of the seeds to Fort Collins to be preserved for posterity.

"If it weren't for these UN people crawling on their bellies under bushes in Africa," one plant scientist observes, "we could well be in trouble in the future." His fears are echoed by other scientists who believe that many of these vanishing primitive plants contain irreplaceable genetic qualities, such as disease resistance and high protein content. Disease resistance in particular is an element that scientists cherish, mainly because of the growing public antipathy to the use of fungicides and pesticides. Even more important is the ever-present danger that a highly destructive disease might overtake a crop—as happened with the disastrous corn blight in 1970. Scientists must then go back to the greenhouse to "design" a variety that will resist the disease. Such plant engineering requires a huge reservoir of genetic material in the form of many different plant varieties.

By any scale, the seed collection at Fort Collins is huge. More than 200 different kinds of plants and 80,-000 separate varieties have been stored away (this includes 7,000 wheat varieties and 9,000 sorghums). Most of these seeds can live 10 to 20 years—and in the case of some tomato seeds the lifespan is over 100 years.

But Mr. Bass can't risk losing a single seed. So from each of the tin cans used for storage, a sample is taken for chemical testing at least every five years. If the seeds have deteriorated, a plant grower is hired to produce new seeds.

Mr. Bass says that his seeds are almost a history of the U.S. in themselves. He has Indian beans and corn collected from reservations and rare strains of native plants, such as the thorn apple and evening primrose; there is a variety of corn grown by the Pilgrim fathers as well as early varieties of cotton; and there are the seeds of delicacies favored by Victorian housewives, such as the speckled cranberry and the vegetable oyster.

Inevitably the laboratory is becoming known as something of a museum. In 1971, for example, agriculture students at the Virginia Polytechnic Institute were assigned the task of growing Eureka and Pamunky corn, two ancient varieties, as part of the college's centennial celebration. The laboratory supplied the only seeds known to exist.

It's not unusual for laboratory employes to start wondering how some of those stored seeds would appear

if allowed to grow. A visitor to the laboratory inquired
about a macadamia-nut tree sprouting gracefully out of
a pot. "We just wanted to see what it looked like," Mr.
Bass said.

—DAVID BRAND

In Search of Superbean

The plant that Richard Cooper holds so lovingly isn't much to look at—just a brittle brown stem with clusters of ripe bean pods strung along it. But this homely plant has taken Mr. Cooper five laborious years of research to produce.

The plant is a dwarf soybean, and from it may spring the first generation of super-productive soybeans. At the Agriculture Department's Regional Soybean Laboratory in Urbana, Ill., where he is chief researcher, Mr. Cooper has discovered that diminutive soybeans are more bountiful than normal four-foot-high soybeans.

Increasing soybean productivity may not sound very startling, but agricultural scientists see this research as a major hope of providing enough protein for a burgeoning world population. The soybean is higher in protein than any other plant known to man, but yields have remained frustratingly low. This has led to a tremendous global demand—and high prices—for the beans, particularly in the U.S., Europe and Japan, where soybeans are the principal high-protein feed for poultry and livestock.

Consumers also are affected directly by soybean production. Soybean oil is used in such products as margarine and shortening, and some Americans are even eating the bean itself as artificial meat, mainly in the form of hamburger "stretcher." For millions of Asians the bean—as a protein curd called tofu—has long been an integral part of a largely vegetable diet.

The search for higher yields is leading scientists into some very odd areas. They have dwarfed the bean plant, emasculated it and gassed it with carbon dioxide. Now they are attempting to "blind" it as well.

Most of these projects probably won't pay off for several years. During that time the world demand for soybeans is expected to soar. Global consumption is increasing by 7% a year, faster than the beans can be grown in the U.S., which provides 90% of the world's exports. (In 1973, these exports—soybeans and their oil and feed products—earned the U.S. $3.84 billion.) However, between 1950 and 1973 U.S. soybean yields increased only six bushels to an average of 27.8 bushels an acre. (By contrast, through complex genetic manipulation U.S. corn yields have been raised in the same 23 years by more than 53 bushels to 91.4 bushels an acre.)

Total U.S. soybean production has soared 423% since 1950, but this has been achieved by quadrupling soybean acreage. The U.S. is fast running out of available cropland, however.

The failure to increase soybean yields is beginning to result in "a deteriorating world supply and a great deal of worry about future protein supplies," says Lester Brown of the Overseas Development Council, a Washington-based think tank. "I doubt whether there is a greater global research priority than this."

Trouble is, the soybean is a most unhelpful plant. It will thrive and produce pods laden with beans (which are the plant's seeds) only when there is the right amount of moisture, warmth and, most of all, daylight. As a result, more than 90% of the world's soybeans are grown in 30 U.S. states, Brazil and China. Varieties have been developed for other areas of the world, such as the tropics and Europe, but these are generally low-yielding.

Agricultural researchers say there isn't a single explanation of why the soybean has so far resisted yield breakthroughs. The soybean's high protein content (about 40% of its weight, compared to 8% for corn) is partly responsible because it takes more solar energy to produce a gram of protein than to produce a gram of carbohydrate.

Mr. Cooper, the Urbana researcher, has found another reason: Soybeans aren't very productive because they're too tall. He discovered that during the early stages of development, when the seed pods are growing, the plant becomes so tall (at least 50 inches high) that it can't support its own weight and bends over. Its thick foliage becomes entangled with the leaves of adjoining soybeans. This blocks out the sunlight from the droopy plant, causing flowers and pods to drop off and sharply reducing potential yields.

Mr. Cooper stumbled on this amazingly simple fact quite by chance. In 1967, he recalls, he planted soybeans on an Illinois farm as part of a research project. Instead of flourishing, the plants were stunted by herbicide that had been used on corn in the field the previous year. But to his amazement the stunted soybeans yielded nearly 25% more than normal.

To prove his conclusion that the normal soybean plant is too tall, the following year he erected a maze of grid wires over a field of soybeans. The wires supported the beans and kept them upright. The result was considerably improved yields.

Mr. Cooper theorized that if a soybean only half the normal height could be developed, the plant would hold up under its own weight. He went to work in 1969 and began looking for genetic traits in various soybean varieties that would give a plant stockiness but also the same number of beans as a normal plant. He found these characteristics in southern U.S. soybean varieties that don't keep growing until they are harvested, as do northern varieties, but grow to a certain height and then stop. After five years of crossing southern and northern varieties, he finally obtained a high-yielding dwarf plant.

But, he says, "I need 80-to-100-bushel (an acre) yields before I can prove I've done it." That would be three times the current national average.

Such high yields would also be possible with a hybrid soybean, the result of emasculating the plant. But the complexity of producing hybrid seeds appears almost overwhelming to many researchers.

A hybrid seed is the offspring of two different varieties of a plant. When the two carefully selected parents are crossed they produce an offspring that has "hybrid vigor"—the capacity to produce many more seeds than a normal plant. No one knows exactly what causes it, but it's a major reason that corn yields have increased so greatly over the past two decades.

The soybean is far more complex to hybridize than corn because it is naturally self-pollinating. The soybean flower contains both the male and female parts and the pollen is shed and fertilization occurs before the flower opens. However, an Agriculture Department scientist, Charles Brim, has found a way to breed a soybean plant in which the male part is sterile. Thus, when the flower opens it is still unfertilized.

But the soybean's pollen is too heavy to be blown by the wind; so, scientists must rely on insects to carry the pollen from a fertile plant to a male sterile plant. "I guess we'd have to train lots of bees if we ever wanted to make this work commercially," Mr. Brim says.

There's another problem: Because of genetics, out of every four seeds produced by Mr. Brim's male-sterile parent, only three blossom into plants that will produce beans; the fourth plant is sterile.

Heavy applications of nitrogen fertilizer—another yield boosting technique used on corn—have no effect on soybeans. They have their own fertilizer factories.

Their roots are invaded naturally by swarms of bacteria that nestle in the plant and extract nitrogen from the air in the soil. The bacteria convert the nitrogen to ammonia, which is passed on to the plant for conversion to protein.

Until recently this natural fertilizer system was considered very efficient. But then Du Pont Co. researchers Ralph Hardy and U. D. Havelka discovered that the bacteria start slowing down during the final 30 days of the plant's growing period, resulting in an inadequate supply of nitrogen when the need is greatest. This means that fewer beans are harvested, since nitrogen is essential to their formation. The researchers realized that if the bacteria could be persuaded to keep up their nitrogen production during those final 30 days, more beans would mature and yields would increase.

Scientists know that the bacteria in the plant's roots are kept alive by sugars supplied by the plant. The soybean makes these sugars by absorbing carbon dioxide from the air through its leaves (the process of photosynthesis). But as the plant matures, Messrs. Hardy and Havelka theorized, more of its sugars go to the developing seeds and fewer go to the bacteria. Thus, nitrogen production slows as the bacteria are denied food.

The answer, the researchers believed, was to feed the plant more carbon dioxide so that it would produce more sugars and keep the bacteria nourished. This they proved by surrounding growing soybean plants with sheets of plastic and pumping carbon dioxide into the enclosed area around the soybean leaves. They found the bacteria had provided the plant with more nitrogen in one week than they normally do during the plant's entire 100-day life. These test plants yielded nearly twice as many beans as usual.

Fertilizing plants with carbon dioxide in a field is impractical. Mr. Hardy says Du Pont hopes to develop a chemical that would increase the soybean plant's appetite for carbon dioxide.

Work on the appetite stimulant at Du Pont is fairly well-advanced, it is thought. But the search at other laboratories for ways to "blind" the soybean is only just getting started.

In order to flower and mature properly the soybean must have a certain amount of daylight. If it is grown too far south it gets too much light, flowers too soon and produces few beans. If it is planted too far north it gets too little light and keeps growing until killed by frost. Thus in the U.S., soybean varieties have been developed that will flourish only at certain latitudes.

Agricultural scientists believe it may be possible to develop genetically a soybean that is "day-neutral." If this could be done, says University of Illinois researcher David Whigham, soybeans might be developed that would grow anywhere in the world, including poor tropical countries and Europe. This would reduce the world's reliance on the U.S. for soybean supplies.

The research is still in its infancy, Mr. Whigham says. But there is promise in a recently discovered soybean plant that appears less sensitive to light than other varieties.

—DAVID BRAND

Bypassing the Birds and Bees

Scientists at the Atomic Energy Commission's Brookhaven National Laboratory have developed a technique for producing hybrid plants that bypasses the usual birds-and-bees approach. They say it opens the way for creating totally new types of crops by crossing widely different plant species—currently difficult, if not impossible—and for improving existing varieties.

Biologists and agricultural specialists hailed the Brookhaven development as an important breakthrough when it was announced in August 1972. Ralph W. Richardson Jr., director of natural and environmental sciences at the Rockefeller Foundation, says the new technique for crossing plants is a "remarkable development" with "fantastic potential" for improving the quality and quantity of man's food crops.

And Adrian Srb, a geneticist at the New York State College of Agriculture at Cornell in Ithaca, called it "a first important step" in the development of "test-tube genetics," whereby scientists could improve crops or create new ones by manipulating plant genes—the tiny particles that determine the characteristics of all living things.

At present the creation of hybrid plants involves the tedious process, sometimes taking decades, of transferring male pollen from one species to the female receptacle of another species, planting the seeds that develop and determining if the offspring are better than the parents.

But to maintain the purity of plant species, nature has erected many formidable barriers to interbreeding. Often, pollen from one species can't be made to fertilize the female parts of another, so most crosses aren't possible.

At Brookhaven, research biologists Peter S. Carlson, Harold H. Smith and Rosemarie D. Dearing didn't bother with the plants' sexual reproductive apparatus. Instead, they took ordinary cells—the small units that are the building blocks of most animals and plants—from the leaves of two different types of tobacco. They then combined both tobacco plant cells, producing a "fused cell" that then grew into a mature hybrid with the characteristics of both parent tobacco plants.

By fusing ordinary leaf cells from different species to create a hybrid, they bypassed the natural barriers that prevent the crossing of most plants with each other. "I believe that any species of plant will fuse with any other," Mr. Carlson, a Ph.D. plant geneticist, said. But he notes that years of work remain to perfect the technique, particularly to create the right conditions for the growth of the hybrid once the parent cells have been fused.

Actually, this isn't the first time cells of different species have been fused. (In fact, some scientists have even grown colonies of cells created by fusing human cells with those of insects. Fortunately, an adult human-insect hasn't been created and probably never will be.) But the Brookhaven scientists are the first to develop methods for growing mature plants from fused plant cells.

The Brookhaven technique should cut down the time needed to create hybrids, Mr. Carlson says. The importance of this can be seen from the fact that high-yielding hybrid corn, which accounts for much of the world's corn production, took decades to develop.

Mr. Richardson of the Rockefeller Foundation says that for years scientists have dreamed of mating corn with sorghum to give corn the drought resistant qualities of sorghum. This hasn't been possible with present methods, he says, but might be with the Brookhaven technique.

Another disadvantage of present hybrids is that most can't reproduce, so hybrid seed must be created anew each year from the parent plants. Because of the way they are created, the Brookhaven hybrids will probably all be fertile; that is, they'll produce seeds generation after generation that will grow into the same hybrid, Mr. Carlson says.

This is because hybrids produced via the sexual route end up with only half the genes from each parent, usually not enough to create a useful offspring. The Brookhaven hybrids, on the other hand, have all the genes from both parents—or double the number of each parent—more than enough to create offspring with all characteristics of the first hybrid.

But the greatest advantage is the wide variety of plant matings the new technique makes possible. At a minimum, says Mr. Srb of Cornell, it greatly increases the array of plant materials with which plant breeders can work.

It's unlikely, the scientists say, that the new technique will be used to create science-fiction type plant crosses, such as stringbean trees, or seaweed that bears apples. It's more likely, they say, that the method will be used to endow present crop varieties with increased resistance to disease, insects or drought, or to improve the yield or protein content.

"Perhaps we'll get something useful one time out of a hundred crosses," says Mr. Carlson, "but that's enough."

Here's how the Brookhaven hybridization works: large numbers of leaf cells were removed from the leaves of two distinct tobacco species. Using enzymes (the chemicals that help the body digest food), the cellulose outer walls of the leaf cells were digested away, leaving the living "innards" that contain the plants' genetic materials.

In the test tube, the wall-less cells were then fused. The broth of fused and unfused cells was then placed in a solution containing chemicals that would only support the growth of the fused cells.

The cells reproduced, forming clumps, or colonies, which were placed in another solution devoid of plant hormones which, the scientists knew, would permit only the fused-cell colonies to form rudimentary shoots and leaves, but not the unfused cells that had slipped through.

The small shoots were then grafted onto freshly cut stems of one of the parent tobacco plants, where they grew into mature hybrids. The two parent tobacco species had been previously crossed sexually, so the Brookhaven scientists could compare their hybrid with that one. It compared in every way, they reported, but unlike normal hybrids that usually can't breed true, the Brookhaven hybird produced seeds that grew into true hybrids.

—Barry Kramer

—Postscript—

Experiments in fused-cell hybridization, now called parasexual hybridization, have spread to laboratories elsewhere in the United States and to Canada, Germany, England and France. Canadian researchers recently fused cells from barley and soybean plants, creating a presumed hybrid of the two species; but they haven't been able yet to grow a mature plant.—B.K.

A New Crop?

The 10-foot-tall kenaf plant doesn't provide shade or housing for squirrels, nor is it strong enough to support a tree house. But in at least one respect, the kenaf plant may someday replace the tree. Kenaf (pronounced kuh-NEF) could be the answer to the paper industry's search for fiber sources.

A type of hibiscus that looks a lot like the marijuana plant, kenaf's main attribute is a fiber yield unsurpassed by any tree yet developed. It produces five to seven times more pulp per acre than the pine tree, for example. And it grows to maturity in only 120 days, compared to 20 years for most trees.

At a time when future wood-fiber shortages are being projected, these qualities make kenaf increasingly attractive to the pulp and paper industry. "The world's pulp and paper industry is becoming more and more concerned about its future sources of fibrous raw material," says Joseph E. Atchison, whose New York consulting firm has evaluated nonwood plant fibers for the pulp and paper industry. "There is no question that nonwood fibers will play an increasing role in satisfying fiber requirements," he says.

The search for nonwood fibers stems from the fact that trees just don't grow fast enough. In fact, there probably aren't enough trees to meet the nation's fiber needs much beyond the 1990s, some paper experts say.

Paper consumption, which in 1974 was 640 pounds per person, is expected to climb to more than 1,000 pounds per person a year by the year 2000. U.S. paper and paperboard production was 61.9 million tons in 1973, more than 3 million tons short of demand.

"It's also possible that trees could become uneconomical (for papermaking) in coming years, as demand for high-cost wood products takes priority," says C. E. MacDonald, International Paper Co.'s director of allied operations. This has already begun to happen; only wood scraps are made into paper in order to conserve more of the tree for paneling, furniture and housing products. "Trees are too valuable to grow just for paper," Mr. MacDonald says.

Though fiber shortages are a new worry for the U.S., many sparsely forested countries have always faced this problem. In Europe and Asia, millions of tons of paper are made from such fibers as bamboo, reeds, rags, bagasse (sugar cane) and straw from wheat, rice, oats and barley. Papyrus, possibly the original paper fiber, is still being used in some countries.

The U.S., too, has a number of varieties of nonwood fibers available. Some 70.8 million tons are produced each year, mostly as a by-product of the harvest of other crops. But wood is much more economical for papermaking than these fibers, and only about one million tons are turned into paper each year.

Although paper can be made from almost any fiber, manufacturers have a long list of criteria for a commercially suitable material. The industry wants a long, strong fiber that can be easily extracted for pulping. It must be versatile and easily handled. And it should be a fast-growing crop that can be easily harvested several times a year. To conserve land and ensure continuous operations, it should have large yields per acre.

With these specifications in mind, scientists have singled out kenaf as a likely possibility to replace wood pulp for papermaking. It was selected after a long investigation by the U.S. Department of Agriculture that began in 1957. The agency started out looking for a good cash crop for farmers and, recognizing the possibility of future fiber shortages, began tests to determine which plant would best fit the paper industry's requirements.

In the course of their investigation, the researchers tested some 6,000 seeds from 3,500 species of fiber plants. Of this group, about 850 were chosen for careful evaluation. Cornstalks, milkweed, honeysuckle and certain types of parsley all showed promise, but were rejected for a variety of reasons. Hemp had strong potential, except that its cultivation is rigidly controlled to prevent the illegal use of its dried leaves as marijuana. Zinnias and sunflowers were deemed too difficult to harvest.

In 1960, kenaf was selected as the best possible candidate—for its strength, harvesting potential, similarity to wood fiber and its pulping and mixing characteristics. "It makes good paper," says Thomas Clark, a USDA kenaf researcher. Mr. Clark, who oversees an experimental kenaf crop at the agency's research laboratory in Peoria, Ill., says the fiber can be made into anything from newsprint to boxes.

But the widespread use of kenaf may still be a long way off, for the paper industry hasn't yet given up on trees. It is devoting huge amounts of time and money to a research effort designed to forestall the wood-fiber shortage.

Manufacturers are turning their attention to trees once considered unsuitable for papermaking; tropical hardwoods such as mahogany are being used for the first time, for example. The increased use of Southern pine trees has relocated the paper industry's operations from the forests of the North to the Plantations of the South.

New scientific developments now make it possible for the eucalyptus tree to grow to 85 feet in five years or less under optimal conditions. And International Paper

Co. has developed fast-growing, sturdy "supertrees." One was nicknamed "Marilyn Monroe" because it was "gorgeous" and had great limbs.

Despite kenaf's high yield and fast growth, it has some drawbacks that may slow its acceptance by paper manufacturers. Like many crops, kenaf must be rotated every few years. Its seedlings rot when exposed to too much water and must be planted in raised beds. The plant would also require an entirely new system of land management. "Paper companies would have to become farmers," says one paper company executive.

The most troublesome problem that must be ironed out before kenaf paper becomes a reality is the root-knot nematode. This nearly microscopic worm burrows into the roots of the plant and saps its growth potential until it wilts. "Unless the bugs are out of it, kenaf will never be economical enough to use," one paper-company executive says.

Though many papermakers share his skepticism about the chances of overcoming this problem, USDA plant geneticist Charles Adamson is more optimistic. He has been trying to develop a nematode-resistant kenaf plant for five years at the agency's Plant Introduction Station in Savannah, Ga. Eventually, he believes, scientists will lick the bug problem.

While kenaf won't be used for tomorrow's newspaper, "it could become a useful source of paper production in the future," according to a spokesman for Weyerhaeuser Co., the timber-products concern. "Kenaf is an excellent fiber," he adds. Gordon B. Killinger, an agronomist at the University of Florida in Gainesville is more emphatic: "Kenaf could be the biggest thing since we learned how to make paper from pine trees back in 1928."

—PAMELA G. HOLLIE

Part Eight

CLOUDS ON THE HORIZON

The very essence of agriculture is optimism. The planting of a seed, the birth of a lamb or a calf, the coming of spring and the bounty of fall—the cycle has repeated itself so many times that it is unthinkable, despite drought or devastation, that it may not always be thus. But the past decade or so has shown that we cannot always continue as we have without paying some painful price sooner or later. It would be nice to think that in the case of agriculture the sky is the limit on increasing yields and productivity. But some people, uneasily eyeing a steadily growing world population, are beginning to think otherwise.

Land of Plenty?

How much longer can U.S. agriculture keep expanding output to provide enough food for this country, let alone to help feed the world?

It seems a strange question to ask in a nation so bountifully blessed that even the farm boom of 1973 wasn't big enough to tax its agricultural capabilities. There are signs, though, that limits are closing in on continued leaping gains in U.S. farm efficiency, which has increased more than sixfold since 1930.

These signs don't necessarily mean the U.S. inevitably will run out of food. Rather, they suggest that some major and probably painful social, economic and political adjustments may be needed to keep our larders stocked as amply as we would like.

The signs are being given more attention now partly because of the growing realizations in recent years that the U.S. doesn't have an inexhaustible supply of natural resources and that unchecked growth isn't always desirable. Also, there is increased concern that next time food shortages throughout the world could develop even more suddenly and severely than they did in the early 1970s, and the U.S. might find itself unable to respond as readily.

Agricultural economists figure that U.S. farms will have to increase production of food and fiber by 32% to meet demand in 1985. Some of the increase will be necessitated by sheer population growth in America and abroad. But rising affluence in the U.S. and among importers of U.S. farm products is also a major factor. As affluence rises, so does consumption of farm products. As people get more money they tend not just to eat more but also to eat more meat, which in turn requires that another cycle be added to farm production—instead of eating grain themselves, people feed it to cattle and then eat the cattle.

Lester R. Brown, senior fellow of the Overseas Development Council, estimates that in recent years rising affluence has increased the global demand for food by half again as much as population growth.

Other experts calculate that there would be little difficulty in expanding U.S. farm output by 20% to 25% in the next several years. If crop prices stay above average, they figure, enough additional land will come into production to meet demand and build up reserve stocks.

But that would still leave projected production lagging behind predicted 1985 consumption by a potential 7% to 12%. This gap could be closed in a number of ways—bringing even more land into production, farming more intensely on acres already under cultivation, striving for some technological breakthrough that would increase yields spectacularly. However, there are pitfalls in all these approaches that could work against ever-increasing abundance.

One of them is that the U.S. is slowly but steadily running out of land that can be easily converted to farm production. There's no real shortage yet—the U.S. has 1.3 billion acres in total, of which well over half is available for agricultural purposes, including livestock grazing. But each year more than one million acres are covered with housing subdivisions, shopping centers, highways and other forms of urbanization, and another two million are given over to recreation and wildlife.

A few states and several localities have become so concerned about the disappearance of farmland that they have passed laws discouraging its commercial development. For instance, Black Hawk County, Iowa, passed a zoning ordinance in 1973 that required at least

35 acres for a homesite to be built on prime agricultural land, thus effectively blocking subdivisions. The ordinance greatly upset the county's farmland owners, who liked the high prices that developers were paying.

In some other areas of the country, less-desirable farm land is being brought into production. The use of irrigation is spreading in the arid Western states, for instance, and in the Southeast old cotton fields and timberland are being cleared and planted with grass for cattle grazing. But these methods are expensive and won't continue to any appreciable extent without sufficient economic incentive—either consistently higher food prices or government subsidies, both of which are unpalatable to consumers and taxpayers.

A few experts think that eventually some segments of U.S. agriculture will lose out in the land squeeze. For instance, D. Gale Johnson, an economist at the University of Chicago, believes the U.S. dairy industry couldn't compete with foreign dairymen if import quotas and other trade barriers were removed. If that happens, Mr. Johnson suggests that dairymen who don't quit altogether could switch to beef raising because in a decade or so the beef industry may be running out of land to expand herds.

Such shifts would be bound to cause severe economic and social upheavals, not to mention the political fights involved in deciding whether this acre or that is to be paved or plowed. "Land-use management will probably cause as much chaos and controversy in coming years as any issue we've had," predicts John M. Trotman, who in 1973 was president of the American National Cattlemen's Association.

Not far behind in controversy will be the growing conflict between environmentalists and agriculturists. Much of the productivity growth in U.S. agriculture has come from farming fewer acres more intensely, and continuing this trend is obviously one way to counter the disappearance of farmland. But intensive farming entails the heavy use of fertilizers and chemicals as well as other practices that are coming under increased scrutiny and challenge.

For instance, officials in some states advocate limiting the amounts and types of fertilizers that farmers can apply because of runoffs of the chemicals into streams and rivers. DDT and other pesticides have been banned or restricted. Some areas are imposing strict water-pollution controls on cattle feedlots and other farms with high concentrations of livestock: One Colorado feedlot had to spend more than $100,000 to comply with the regulations or face being shut down.

Thus far, the environmentalists seem to have won the most battles, but farm leaders are using the food shortages that popped up in 1973 to bolster their side of the argument. In an article entitled "Roadblocks to Abundance," American Farm Bureau Federation President William J. Kuhfuss accused "the Environmental Protection Agency (of) being more concerned with emotion and trivia than with reality." Observers expect this conflict to heat up in coming years, with the possibility that many of the former means of boosting production could be eliminated or curtailed.

A third avenue for expanding farm output is through improved technology. The development of hybrid corn in the 1930s was probably the most dramatic breakthrough in agriculture this century, because it laid the groundwork for the rapid expansion of meat production in the 1950s and 1960s. Researchers in government, universities and corporations are busily trying for other breakthroughs, but progress is slow and uncertain.

Research on hybrid wheat, for example, has been underway for more than 20 years and has cost several million dollars, but the hybrid is still in the experimental stage and won't be generally available until about 1978 or so; even then, it is expected to boost yields by only 20%-25%, compared to more than 40% for hybrid corn when it was introduced.

Hybrid soybeans, multiple calf births and other possible means of substantially boosting output are much more remote, some scientists say. The government, in an effort to trim farm spending, cut agriculture research funds by almost 9% in fiscal 1974; thus possibly prolonging the search for technological breakthroughs.

Meanwhile, more immediate hurdles are impeding production increases. Among them are:

—Fertilizer shortages, which experts say could last until 1976 or later, especially for fertilizers containing nitrogen and phosphates. Fertilizers are necessary not only to boost yields but to bring retired acreage back to full strength for economic production.

—Fuel shortages, which affect farmers more and more as they rely increasingly on bigger machines and on grain drying to keep their harvests from rotting. The national allocation program gives farmers high priority —as they had during World War II—but farmers could be stymied by either shortages or high prices in their rapid expansion of fuel consumption, which in turn could dampen their enthusiasm for boosting output. Tight natural-gas supplies also are holding down production of ammonia fertilizer, which is in great demand in many farm areas.

—Rail-car shortages, which slow the flow of supplies—such as fertilizer—to farmers and the movement of their produce to market. Railroads are adding more grain hopper cars and doing other things to speed service, but not as much as farmers and food companies would like. "The rail-car shortage is feeding on itself," says Sam H. Flint, vice president of corporate operations for Quaker Oats Co. "Prospects for an early solution are not bright."

Small wonder, then, that many observers think U.S. agriculture is entering an era more of knotty problems than of unbounded growth. "Increased productivity is the only way to halt the rise in food prices for the consumer even while maintaining the incentive of the American farmer to produce the needed crops and livestock products," says James W. Hogan, former president of the National Soybean Processors Association and vice president of Ralston Purina Co. "But with limits on acreage available for expansion, the difficulties in developing hybrids, and other problems, it's not going to be as easy as we might wish."

—NORMAN H. FISCHER